Ocean Wave Data Analysis

Introduction to Time Series Analysis, Signal Processing, and Wave Prediction

Arash Karimpour

ISBN-13: 978-0-692-10997-7

"… and we made from water every living thing …" [21:30]

To my parents, my wife, and my family

Preface

This textbook helps readers to better understand the process involved in the ocean wave measurement, analysis, and prediction. Although there is a rich literature about a time series analysis and signal processing in general, there are not as many options available when it comes to ocean wave analysis. This book intends to fill this gap and provides an introduction to time series analysis, signal processing, and prediction of the ocean wave data.

Ocean wave studies mostly require some sort of time series analysis. To achieve that, a time series of the wave data are commonly analyzed and processed in one of the time or frequency domain. The outcome of this analysis helps to calculate wave properties representing a wave dataset and can provide a comprehensive perspective of the wave field.

In this book, readers are walked through steps that are required for the ocean wave data analysis. This book covers concepts and theories of the time series analysis, signal processing, and ocean wave prediction in simple language. This text explains how to implement these theories for a purpose of the ocean wave data analysis. This book is prepared with a practical application in mind. Throughout this book, it is explained how to achieve an intended goal for each section, and how to calculate associated parameters described in that section.

The first section of this book provides an introduction to data analysis in the time and frequency domains. The second section of this text focuses on how to analyze measured wave data. The third section of the book describes how to predict, estimate, and model ocean waves. Codes written in MATLAB, GNU Octave, and Python are provided throughout the book to calculate different wave properties.

Contents

1 Introduction to Wave Theory

In this chapter, a basic of the wave theory is covered. First, the linear wave theory is explained, and then higher order wave theories are discussed.

1.1 Linear wave theory

Gravity waves are results of the water displacement under external factors such as wind, or internal factors such as density differences. In this process, the gravity and buoyancy play as equilibrium restoring forces. Gravity waves that are generated at an interface of the air and water media are called surface gravity waves or surface waves. Gravity waves that are generated inside a water media are called internal waves.

The linear wave theory, the so-called small amplitude wave theory or Airy wave theory, describes generation and propagation of the surface gravity waves, such as wind-generated waves. The linear wave theory is only applicable to a sinusoidal wave with a small wave height compared to its wavelength and water depth.

1.2 Basic relations

In the linear wave theory, propagation of the surface sinusoidal wave is described as:

$$\eta = a\cos(kx - \omega t + \psi) = \frac{H}{2}\cos\left(\frac{2\pi}{L}x - \frac{2\pi}{T}t + \phi\right) \tag{1.1}$$

$$\omega = \frac{2\pi}{T} \tag{1.2}$$

$$k = \frac{2\pi}{L} \tag{1.3}$$

$$T = \frac{1}{f} \tag{1.4}$$

where, η is a water surface elevation known as an excursion, $a = H/2$ is an amplitude, $\omega = 2\pi/T$ is an angular frequency, $k = 2\pi/L$ is a wavenumber, x is a distance, t is time, $(kx - \omega t)$ is a wave phase in $radian$, ϕ is a random phase angle or a phase shift in $radan$, which commonly is considered as $\phi = 0$, H is a wave height, $T = 1/f$ is a wave period, L is a wavelength, and f is a wave frequency in Hz.

Above parameters describe wave properties in time and space domains. In the time domain, a wave period, T, defines the time that wave completes one full oscillation. A wave frequency, f, with a unit of Hertz, Hz, represents a number of full oscillations per unit time. Similar to a frequency, an angular frequency, ω, with a unit of $radian/s$ shows an angular displacement per unit time.

In the space domain, a wave height, H, which is two times of the wave amplitude, a, is equal to a vertical distance between a wave crest and a neighboring trough. A wavelength, L, shows a distance that a wave travels in one full oscillation. In other words, it is a distance between two consecutive similar points such as two consecutive crests or two consecutive troughs. A wavenumber, k, with a unit of $radian/m$, which is similar to an angular frequency, represents radians per unit distance or cycles per unit distance.

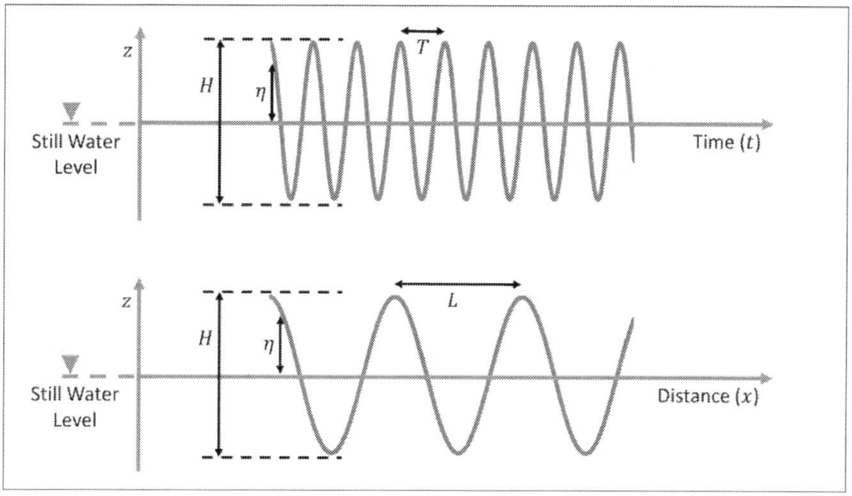

Figure 1.1 Demonstration of the wave parameters in the time and space domain

1.3 Wave characteristics based on water depth

Waves are categorized based on water depth and a wavelength as:

- Shallow water waves: where $h/L \leq 0.05$ or $\tanh(kh) \approx kh$
- Intermediate water waves: where $0.05 \leq h/L \leq 0.5$
- Deep water waves: where $h/L \geq 0.5$ or $\tanh(kh) \approx 1$

where, h is water depth. Deep water parameters are commonly demonstrated by subscript 0. Among wave parameters, a wave period is assumed to stay constant when a wave travels from deep water to shallow water.

1.4 Dispersion relation

A dispersion relationship links the space domain with the time domain by relating a wavelength or a wavenumber with a wave frequency as:

$$\omega^2 = gk \tanh(kh) \tag{1.5}$$

A dispersion relationship shows that waves with different wavelengths travel with different phase speeds. For a given depth, longer waves propagate faster. For a given wavelength, waves have larger phase speeds in deeper water.

If waves travel on top of the moving water with a mean flow velocity of U, then waves experience a Doppler shift. A dispersion relation for a moving water body is:

$$\omega^2 = gk\tanh(kh) + kU \tag{1.6}$$

where, U is mean water velocity. An angular frequency calculated from $\omega^2 = gk\tanh(kh)$ is called an intrinsic angular frequency, which also is represented by $\sigma = \sqrt{gk\tanh(kh)}$. An angular frequency calculated from $\omega^2 = gk\tanh(kh) + kU$ is known as an apparent angular frequency.

1.5 Wave phase velocity (Celerity)

Phase velocity shows a wave phase propagation rate in space. In other words, phase velocity, a so-called celerity, shows at what velocity wave travels in space. Any phase of the wave, such a crest or trough, travels at phase velocity. Phase velocity is presented as:

$$C = \frac{L}{T} = \frac{\omega}{k} = \frac{g}{\omega}\tanh(kh) = \frac{gT}{2\pi}\tanh(kh) = \sqrt{\frac{g}{k}\tanh(kh)} \tag{1.7}$$

where, C is a wave celerity or wave phase velocity. Phase velocity also can be presented in a term of the relative depth, h/L, as:

$$C = \frac{g}{\omega}kh = \frac{gTh}{L} = \sqrt{gh} \quad \text{for shallow water, where } \tanh(kh) \approx kh \tag{1.8}$$

$$C = \frac{g}{\omega}\tanh(kh) = \frac{gT}{2\pi}\tanh(kh) = \sqrt{\frac{g}{k}\tanh(kh)} \quad \text{for intermediate water} \tag{1.9}$$

$$C_0 = \frac{g}{\omega_0} = \frac{gT}{2\pi} = \sqrt{\frac{g}{k_0}} \quad \text{for deep water, where } \tanh(kh) \approx 1 \tag{1.10}$$

where, g is the gravitational acceleration. Subscript 0 indicates a deep-water condition.

1.6 Group wave velocity

A group velocity shows a velocity that an overall shape of the wave amplitudes travels through space. A group velocity, also known as a modulation or an envelope of the wave, is defined as:

$$C_g = C \quad \text{for shallow water, where } h/L \leq 0.05 \text{ or } \tanh(kh) \approx kh \tag{1.11}$$

$$C_g = \frac{C}{2}\left(1 + \frac{2kh}{\sinh(2kh)}\right) \quad \text{for intermediate water, where } 0.05 \leq h/L \leq 0.5 \tag{1.12}$$

$$C_g = \frac{C}{2} \quad \text{for deep water, where } h/L \geq 0.5 \text{ or } \tanh(kh) \approx 1 \tag{1.13}$$

where, C_g is a group wave velocity.

1.7 Water particle motion under the wave

Water particle motion or an excursion under a wave is expressed as:

$$\zeta = -\frac{H}{2}\frac{\cosh(k(z+h))}{\sinh(kh)}\sin(kx - \omega t) \tag{1.14}$$

$$\xi = \frac{H}{2}\frac{\sinh(k(z+h))}{\sinh(kh)}\cos(kx - \omega t) \tag{1.15}$$

Water particle velocities under a wave are:

$$u = \frac{\pi H}{T}\frac{\cosh(k(z+h))}{\sinh(kh)}\cos(kx - \omega t) \tag{1.16}$$

$$w = \frac{\pi H}{T}\frac{\sinh(k(z+h))}{\sinh(kh)}\sin(kx - \omega t) \tag{1.17}$$

Water particle accelerations under a wave are defined by:

$$a_x = \frac{2\pi^2 H}{T^2}\frac{\cosh(k(z+h))}{\sinh(kh)}\sin(kx - \omega t) \tag{1.18}$$

$$a_z = -\frac{2\pi^2 H}{T^2}\frac{\sinh(k(z+h))}{\sinh(kh)}\cos(kx - \omega t) \tag{1.19}$$

A maximum near bed orbital velocity is:

$$u_m = \frac{\pi H}{T\sinh(kh)} \tag{1.20}$$

where, ζ and ξ are displacements of the water particles in x (horizontal) and z (vertical) directions, respectively. The u and w are water particle velocities in x and z directions, respectively. The a_x and a_z are water particle accelerations in x and z directions, respectively. The z is an upward vertical axis with zero at a water surface. In other words, z is an upward distance in a vertical direction with $z = 0$ at a water surface. The z is positive above a water surface and is negative under a water surface. The h is water depth. Note that, water depth is measured from a water surface downward, and therefore, all depth values are positive. The u_m is a maximum near bed orbital velocity.

1.8 Water pressure under the wave

Water pressure under a wave consists of two components. The first one is a hydrostatic pressure, which is a result of the still water pressure and represents water depth. The second one is a dynamic pressure, which is a result of the water surface fluctuations and represents wave activities. Therefore, the total pressure under a wave is presented as:

$$P_0 = P + q = \rho g h + \rho g \eta \frac{\cosh(k(h+z))}{\cosh(kh)} = \rho g h + \rho g \eta K_P \tag{1.21}$$

$$K_p = \frac{\cosh(k(h+z))}{\cosh(kh)} \tag{1.22}$$

where, P_0 is the total pressure, P is a static pressure, q is a dynamic pressure, K_p is a dynamic pressure to a water surface elevation conversion factor, also known as a pressure response factor or a pressure attenuation factor, and z is an upward vertical axis with zero at a water surface.

A dynamic component of the pressure, which is resulted from a wave activity, attenuates in a water column by traveling from a water surface toward a bed. A pressure response factor, K_P, represents this attenuation. A pressure response factor is equal to one at a water surface, i.e. $K_P = 1$ at $z = 0$. It starts to decrease as water depth increases until it becomes $K_P \approx 0$ at $z \leq -0.5L$. In other words, K_P becomes zero at a depth where a deep-water condition, i.e. $h/L \geq 0.5$, is met.

1.9 Wave energy and power

Wave energy is a sum of the potential and kinetic energies, assuming a surface tension energy is negligible. Potential energy is a result of the water surface fluctuations away from a mean water surface level. Kinetic energy is a result of the water particle orbital motions under waves. Mean values of the potential and kinetic energies within a wavelength of the wave are defined as:

$$E_P L = E_K L = \left(\frac{1}{16}\rho g H^2\right) L \tag{1.23}$$

$$E = E_P + E_K = \frac{1}{8}\rho g H^2 \tag{1.24}$$

where, E_P, E_K and E are mean values of the potential, kinetic and total energy densities per unit horizontal area of the water surface with a unit of joule per unit area, J/m^2, respectively, and ρ is water density.

Wave power, a so-called energy flux, is an energy transfer rate through a surface. As a wave propagates in water, wave energy is transmitted by potential, kinetic and pressure energies of the individual particles. A velocity of the energy transportation is equal to a group velocity. Therefore, wave power or wave energy flux through a unit width of the vertical plane perpendicular to a wave propagation direction is:

$$P_w = EC_g \tag{1.25}$$

where, P_w is wave power with a unit of watt per meter, W/m. An energy transfer rate depends on water depth. In deep water where $C_g = C/2$, wave energy is transmitted with half of the wave celerity, while in shallow water where $C_g = C$, wave energy is transmitted with the same velocity as a wave celerity.

1.10 Wave radiation stress (momentum flux)

Radiation stress is an excess flow of the momentum caused by a presence of the surface waves. It is an extra force due to wave activity, which changes a mean value of the depth integrated horizontal momentum. Radiation stress causes phenomena such as a wave setup and wave-induced currents to happen. It can be calculated from a depth-integrated and a phase-averaged integration of the dynamic pressure.

For shallow water, where $h/L \leq 0.05$ or $\tanh(kh) \approx kh$:

$$S_{XX} = \frac{3}{2}E \tag{1.26}$$

$$S_{YY} = \frac{1}{2}E \tag{1.27}$$

For intermediate water, where $0.05 \leq h/L \leq 0.5$:

$$S_{XX} = E\left(\frac{2kh}{\sinh(2kh)} + \frac{1}{2}\right) = E\left(2\frac{C_g}{C} - \frac{1}{2}\right) \tag{1.28}$$

$$S_{YY} = E\left(\frac{kh}{\sinh(2kh)}\right) = E\left(\frac{C_g}{C} - \frac{1}{2}\right) \tag{1.29}$$

For deep water, where $h/L \geq 0.5$ or $\tanh(kh) \approx 1$:

$$S_{XX} = \frac{1}{2}E \tag{1.30}$$

Table 1.1 Linear wave theory

Parameter	Variable	Shallow Water	Intermediate Water	Deep Water
Relative water depth	h/L	$h/L \leq 0.05$	$0.05 \leq h/L \leq 0.5$	$h_0/L_0 \geq 0.5$
		$\tanh(kh) \approx kh$		$\tanh(k_0 h_0) \approx 1$
Wave profile	η	$\eta = a\cos(kx - \omega t)$	$\eta = a\cos(kx - \omega t)$	$\eta = a\cos(kx - \omega t)$
Dispersion relationship		$\omega^2 = gk^2 h$	$\omega^2 = gk\tanh(kh)$	$\omega_0^2 = gk$
Phase velocity (celerity)	c	$C = \frac{g}{\omega}kh$	$C = \frac{g}{\omega}\tanh(kh)$	$C_0 = g/\omega_0$
		$C = \frac{gTh}{L}$	$C = \frac{gT}{2\pi}\tanh(kh)$	$C_0 = \frac{gT}{2\pi}$
		$C = \sqrt{gh}$	$C = \sqrt{\frac{g}{k}\tanh(kh)}$	$C_0 = \sqrt{\frac{g}{k}}$
Group velocity	c_g	$C_g = C$	$C_g = \frac{C}{2}\left(1 + \frac{2kh}{\sinh(2kh)}\right)$	$C_g = \frac{C_0}{2}$
Radiation stress in x direction	S_{XX}	$S_{XX} = \frac{3}{2}E$	$S_{XX} = E\left(\frac{2kh}{\sinh(2kh)} + \frac{1}{2}\right)$	$S_{XX} = \frac{1}{2}E$
			$S_{XX} = E\left(2\frac{C_g}{C} - \frac{1}{2}\right)$	
Radiation stress in y direction	S_{YY}	$S_{YY} = \frac{1}{2}E$	$S_{YY} = E\left(\frac{kh}{\sinh(2kh)}\right)$	$S_{YY} = 0$
			$S_{YY} = E\left(\frac{C_g}{C} - \frac{1}{2}\right)$	

$$S_{YY} = 0 \tag{1.31}$$

where, S_{XX} and S_{YY} are radiation stresses in x and y directions with a unit of force per unit of length, N/m, respectively.

1.11 Higher order wave theory

The linear wave theory, a so-called small-amplitude wave theory, is appropriate for a sinusoidal wave with a small wave steepness, i.e. $H/L \ll 1$, and a small wave height to water depth ratio, i.e. $H/h \ll 1$. In other words, it is applicable only to a sinusoidal wave with a small wave height compared to its wavelength and water depth. The small-amplitude wave theory is appropriate for a full range of the relative water depth, h/L, but it is only appropriate for waves with small H/h in shallow water and with small H/L in deep water.

Note that, the mostly used limit for wave growth before it breaks is $H_b/L_b = 1/7$ in deep water and is $H_b/h_b = 0.78$ in shallow water, where subscript b indicates a breaking condition.

For conditions that the small-amplitude wave theory is not appropriate, the finite-amplitude wave theory is considered. Unlike the small-amplitude wave theory which is valid if both H/L and H/h are small, the finite-amplitude wave theory was developed for a condition where only one of H/L or H/h has a small value. A finite wave steepness, H/L, leads to relationships that are more suitable for deep and intermediate water waves, while a finite H/h results in relationships that are more suitable for shallow water waves (e.g. Sorensen, 2005).

1.12 Stokes waves

Stokes' wave is a non-linear wave theory which is appropriate for steep waves with a finite steepness of $H/L \geq 0.04$. It is used for waves with small H/h in deep and part of intermediate water where $h/L > 0.1$. A surface profile under second-order Stokes' wave is:

$$\eta = a\cos(kx - \omega t) + ka^2 \frac{\cosh(kh)}{4\sinh^3(kh)}(2 + \cosh(2kh))\cos\big(2(kx - \omega t)\big) \tag{1.32}$$

$$\eta = \frac{H}{2}\cos\left(\frac{2\pi}{L}x - \frac{2\pi}{T}t\right) + \frac{\pi H}{8}\left(\frac{H}{L}\right)\frac{\cosh(kh)}{\sinh^3(kh)}(2 + \cosh(2kh))\cos\left(2\left(\frac{2\pi}{L}x - \frac{2\pi}{T}t\right)\right) \tag{1.33}$$

Then, a surface profile under the second-order Stokes' wave in deep water is:

$$\eta = \frac{H_0}{2}\cos\left(\frac{2\pi}{L}x - \frac{2\pi}{T}t\right) + \frac{\pi H_0}{4}\left(\frac{H_0}{L_0}\right)\cos\left(2\left(\frac{2\pi}{L}x - \frac{2\pi}{T}t\right)\right) \tag{1.34}$$

Water particle motion or excursions under a wave is expressed as:

$$\zeta = -\frac{H}{2}\frac{\cosh\big(k(z+h)\big)}{\sinh(kh)}\sin(kx - \omega t) + \frac{\pi H^2}{8L\sinh^2(kh)}\left(1 - \frac{3\cosh\big(2k(z+h)\big)}{2\sinh^2(kh)}\right)\sin\big(2(kx - \omega t)\big) + \frac{\pi H^2}{4L}\frac{\cosh\big(2k(z+h)\big)}{\sinh^2(kh)}(\omega t) \tag{1.35}$$

$$\xi = \frac{H}{2}\frac{\sinh(k(z+h))}{\sinh(kh)}\cos(kx-\omega t) + \frac{3\pi H^2}{16L}\frac{\sinh(2k(z+h))}{\sinh^4(kh)}\cos(2(kx-\omega t)) \quad (1.36)$$

Water particle velocities under the second-order Stokes' wave are:

$$u = \frac{\pi H}{T}\frac{\cosh(k(z+h))}{\sinh(kh)}\cos(kx-\omega t) + \frac{3(\pi H)^2}{4TL}\frac{\cosh(2k(z+h))}{\sinh^4(kh)}\cos(2(kx-\omega t)) \quad (1.37)$$

$$w = \frac{\pi H}{T}\frac{\sinh(k(z+h))}{\sinh(kh)}\sin(kx-\omega t) + \frac{3(\pi H)^2}{4TL}\frac{\sinh(2k(z+h))}{\sinh^4(kh)}\sin(2(kx-\omega t)) \quad (1.38)$$

Water particle accelerations under the second-order Stokes' wave are:

$$a_x = \frac{2\pi^2 H}{T^2}\frac{\cosh(k(z+h))}{\sinh(kh)}\sin(kx-\omega t) + \frac{3\pi^3 H^2}{T^2 L}\frac{\cosh(2k(z+h))}{\sinh^4(kh)}\sin(2(kx-\omega t)) \quad (1.39)$$

$$a_z = -\frac{2\pi^2 H}{T^2}\frac{\sinh(k(z+h))}{\sinh(kh)}\cos(kx-\omega t) - \frac{3\pi^3 H^2}{T^2 L}\frac{\sinh(2k(z+h))}{\sinh^4(kh)}\cos(2(kx-\omega t)) \quad (1.40)$$

A dispersion relation for the third-order Stokes' wave is:

$$C^2 = \frac{g}{k}\tanh(kh)\left(1 + \left(\frac{\pi H}{L}\right)^2\left(\frac{9+8\cosh^4(kh)-8\sinh^2(kh)}{8\sinh^4(kh)}\right)\right) \quad (1.41)$$

A wave celerity under the Stokes' wave is:

$$C = \frac{gT}{2\pi}\tanh(kh)\left(1 + \left(\frac{\pi H}{L}\right)^2\left(\frac{5+2\cosh(2kh)+2\cosh^2(2kh)}{8\sinh^4(kh)}\right)\right) \quad (1.42)$$

Then, a wave celerity under the Stokes' wave in deep water is:

$$C_0 = \sqrt{\frac{gL_0}{2\pi}\left(1 + \left(\frac{\pi H_0}{L_0}\right)^2\right)} \quad (1.43)$$

Subsurface pressure under the Stokes' wave is:

$$P_0 = \rho g \frac{H}{2}\frac{\cosh(k(h+z))}{\cosh(kh)}\cos(kx - \omega t) + \frac{3}{8}\rho g \frac{\pi H^2}{L}\frac{\tanh(kh)}{\sinh^2(kh)}\left(\frac{\cosh(2k(h+z))}{\sinh^2(kh)} - \frac{1}{3}\right)\cos\big(2(kx - \omega t)\big) - \rho g z - \frac{1}{8}\rho g \frac{\pi H^2}{L}\frac{\tanh(kh)}{\sinh^2(kh)}\big(\cosh(2k(h+z)) - 1\big) \tag{1.44}$$

A mass transport velocity, also called the Stokes drift, $\bar{u}$, is calculated from:

$$\bar{u} = \left(\frac{\pi H}{L}\right)^2 \frac{C}{2}\frac{\cosh(2k(z+h))}{\sinh^2(kh)} = \frac{\pi^2 H^2}{2TL}\frac{\cosh(2k(z+h))}{\sinh^2(kh)} \tag{1.45}$$

1.13 Cnoidal waves

The Stokes' wave theory is suitable for deep and part of intermediate water. In shallow water, the Cnoidal wave theory needs to be used. The Cnoidal wave theory is appropriate for waves with a finite H/h in shallow water where $h/L < 0.1$.

1.14 Solitary waves

Solitary waves are suitable for very shallow water with decreasing depth. A crest of the solitary wave is above a water surface and it has no trough.

2 Introduction to Time Series: Data Acquisition and Preparation

In this chapter, basics of the time series, data sampling, and data preparation are covered. First, sampling methods are explained, next data quality control is discussed, and then de-trending of the data is discussed.

2.1 Sampling frequency

In coastal engineering and oceanography studies, it is a common practice to explore a phenomenon such as wave activities through in situ measurements and data collection. For the purpose of data collection, data are commonly recorded in the form of the time series. These time series consist of successive data points which are generally measured with an equal time interval between every two consecutive points. By knowing this time interval between points, a sampling frequency of the measurement, f_s, is defined as $f_s = 1/\Delta t_s$, where Δt_s is a time interval between two sequential data points. A value of the sampling frequency, f_s, represents a total number of measured data points in one second of the time series. In other words, there are f_s data points in every second of the measured time series.

To get a reliable measured dataset, a sampling frequency, f_s, should be at least 2 times of the largest frequency that is needed to be studied (the Nyquist frequency). It means that, if data are sampled with a sampling frequency equal to f_s Hz, then only signals with a frequency range of $0\ Hz < f \leq f_s/2\ Hz$ can be studied. For example, if a study of waves with frequencies up to 2 Hz is intended, then a sampling frequency needs to be to be equal or larger than 4 Hz, i.e. $f_s \geq 4\ Hz$. This sampling frequency needs to be defined prior to deployment of the instrument, which means it should be chosen before measurement starts. For the purpose of wave measurements, researchers commonly choose a sampling frequency based on characteristics of the waves in an area of interest. In areas where short waves are present, such as in shallow water, a sampling frequency around $f_s = 8\ Hz$ is a suitable choice. In areas where longer waves are present, such as in deep water, a lower sampling frequency like $f_s = 1\ Hz$ is sufficient. These suggestions are suitable for areas without a considerable current under waves. Choosing a sampling frequency for an area with a considerable current needs additional attention. For choosing a minimum sampling frequency, f_s, that is required for waves sampling in the presence of the current, refer to study by Jones and Monismith (2007).

A sampling frequency, f_s, for wave data collections needs to be chosen carefully. A low sampling frequency results in data loss associated with shorter waves. In other words, if a low sampling frequency is used, then waves with frequencies larger than half of the sampling frequency, i.e. $f > f_s/2$, cannot be detected. For example, if the shortest wave in the area has a frequency of $f = 4\ Hz$ and a sampling frequency is chosen as $f_s = 1\ Hz$, then all data with a frequency range of $f > 1/2\ Hz$ are lost. In contrast, a high sampling frequency results in large memory storage. Choosing an unnecessary large sampling frequency results in collecting unavailable or undesirable high frequency waves, requiring a large amount of storage. For example, if the shortest wave in the area has a frequency of $f = 4\ Hz$ and a sampling frequency is chosen as $f_s = 16\ Hz$, then all collected data with a frequency range of $4\ Hz < f \leq 16/2\ Hz$ are without any analytical value and benefit.

2.2 Data sampling in burst and continuous modes

During a measurement process, data either can be collected continuously, i.e. in a continuous mode, or can be collected in a predefined interval for a specific period (duration) of time, i.e. in a burst mode. In the first method, i.e. in a continuous mode, a sensor keeps sampling at an interval of Δt_s seconds without any break throughout a deployment or measurement period. In the first method, i.e. in the burst mode, an instrument is set to wake up in a time interval of Δt_{bl} and to record the data at an interval of Δt_s for a specific duration of t_{bl}, before it goes back to sleep mode. A time interval between instrument waking up, Δt_{bl}, is mostly set equal to 30 or 60 minutes. Using a continuous measurement is preferable in cases that there is no power or data storage memory restraint, such as in laboratory measurements. Otherwise, if there is a power or data storage memory limitation, such as in field measurements, then a burst mode is more feasible. Setting up sensors in a burst mode arrangement helps to reduce the power and memory consumption and to increase a length of the deployment period.

2.3 Sampling duration

Regardless if data are measured in a continuous or a burst mode, data need to be split into a series of bursts for analysis purposes. In this case, each burst is a segment of data that contains a collection of consecutive data points for a specific duration. If data are recorded in a burst mode, data already are split into a series of data blocks at a Δt_{bl} interval. However, if data are recorded in a continuous mode, then data need to be split into blocks of data at an interval of Δt_{bl}. After blocks of data are defined from either of these methods, then a burst of data within each block is chosen. A total number of data points that are chosen from each block to represent one burst of data should satisfy two conditions. First, a total number of data points in a burst have to be large enough to be a good representative of that block. Second, data points in a burst have to be independent of the time, i.e. data in a burst should be in a stationary state. In other words, each burst has to have enough representing stationary state samples. To ensure these conditions are met, it is recommended that each burst only contains about 15 to 20 minutes of data from each data block. Also, a total number of samples for each burst are

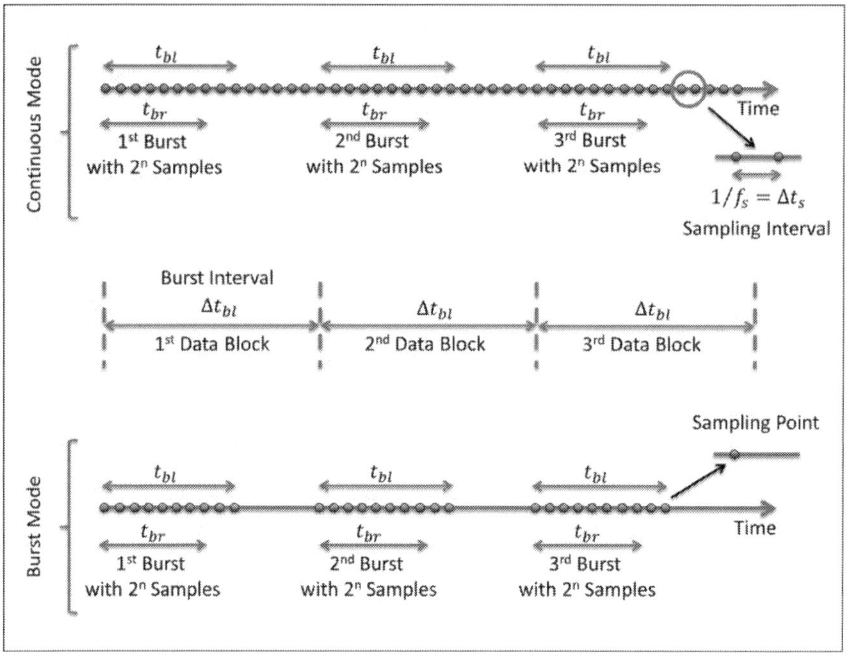

Figure 2.1 Definition of the measurement modes. Top: continuous modes. Bottom: burst mode. Middle: shows how data blocks are chosen in continuous and burst modes.

preferred to be selected in a way that they can be represented by 2^n. In other words, each burst is better to have 2^n samples.

As an example, if data are recorded in a burst mode for a duration of $t_{bl} = 1200\ s$ at an interval of $\Delta t_{bl} = 1\ hr$ with a sampling frequency of $f_s = 4\ Hz$, then there are $1200\ s \times 4\ Hz = 4800$ data points in each $1\ hr$ block of data. Then, each burst of data, which represents a $1\ hr$ block, can be defined with a duration equal to $t_{br} = 1024\ s$, which is about 17 minutes of data. In this case, each burst contains $1024\ s \times 4\ Hz = 4096 = 2^{12}$ samples.

As another example, assume that data are recorded in a continuous mode with a sampling frequency of $f_s = 8\ Hz$. Now, assume that an hourly analysis of data is desired. Therefore, data are split into an hourly block with $\Delta t_{bl} = 1\ hr$. Each block of data, which represents an hour of measurement, contains $3600\ s \times 8\ Hz = 28800$ data points. Then, a burst of data to represent this $1\ hr$ block, is defined with a duration equal to $t_{br} = 1024\ s$, which is about 17 minutes of data. In this case, each burst contains $1024\,s \times 8\,Hz = 8192 = 2^{13}$ samples. In another example, assume that analysis of data is desired in a 30-minute interval. Therefore, data are split into a series of data blocks with $\Delta t_{bl} = 30\,min$. Each 30-min block of data contains $1800\,s \times 8\,Hz = 14400$ data points. Again, a burst of data, which in this case represents a 30-min block of data, is chosen with a duration equal to $t_{br} = 1024\,s$. Similarly, each burst contains $1024\,s \times 8\,Hz = 8192 = 2^{13}$ samples, but representing a 30-min block instead of the 1 hr block. The reason that t_{br} is chosen equal to $t_{br} = 1024\,s$ in all these examples is that $t_{br} = 1024\,s$ is long enough to represent a 30-min or a 1 hr block of data and is short enough to be stationary.

2.4 Data quality control

In any data collection procedure, after data are collected, quality of the recorded data needs to be checked. As the first step, overall values of the measured data are evaluated to be within an accepted range for a measured parameter. For example, assume water depth is measured in a shallow water basin with a depth range of 1 to 2 m. Now, if all measured depths in a dataset are larger than 10 m, then there might be a fault with a sensor or a measurement procedure. After an overall range of the measured data is found to be acceptable, then data are checked for spikes, unacceptable values, and missing values.

2.4.1 de-spiking

Spikes in the data are referred to an unacceptable sudden rise or fall of the data in a time series. These spikes need to be removed from the data. This procedure of removing spike points is known as a de-spiking procedure. To treat the spikes, or in other words to de-spike a dataset, spike values are replaced using adjacent data points. The first method to de-spike dataset is to simply replace a spike value with a value from the nearest neighbor in a time series. The more accurate method is to replace a spike value by interpolating a value at a spike location from adjacent data in a time series. For example, a spike value can be replaced by an average value of two neighbor data points, one right before and one right after a spike location.

2.4.2 Unacceptable data removal

Depending on the nature of the data, some dataset cannot have certain values, such as zero. For example, assume water depth is measured in an area with an average depth 10 m. In this case, any negative or zero value in water depth measurements is not acceptable and needs to be replaced, assuming a depth is considered positive. Similar to a de-spiking method, these unacceptable values can be replaced either by values from the nearest neighbor or by values interpolated from the adjacent data in a time series.

2.4.3 Missing data

Measured data should be evaluated to make sure there is no missing data point in a recording. In case that there are a limited number of missing data points in a dataset, similar to a de-spiking method, these missing points can be replaced using adjacent data points.

Note that, for any of these cases, if there exist a large number of spikes, unacceptable values, or missing values, then data might not be reliable. For example, if there is a significant gap in a recording, then missing data points should be replaced from another source such as a nearby sensor or station. If there are no other sources to help replacing these missing data, then unreliable data should be removed from a dataset. After removing unreliable data, a remaining of the dataset should be evaluated to see if it can be an acceptable representative for a duration of the intended analysis or not. If it is not, then data are not acceptable for an analysis.

2.5 de-trending data

After the quality of data is checked, data in each burst need to be evaluated for an existing trend in a time series. An ideal signal records have an average value equal to zero at any given period. It means that an average value of the ideal signal calculated over any given period of time always stays zero. However, it might not be true for measured signals. Therefore, after data are split into a series of bursts, and before starting a signal analysis, the time series data in each burst should be evaluated for a presence of any trend in a

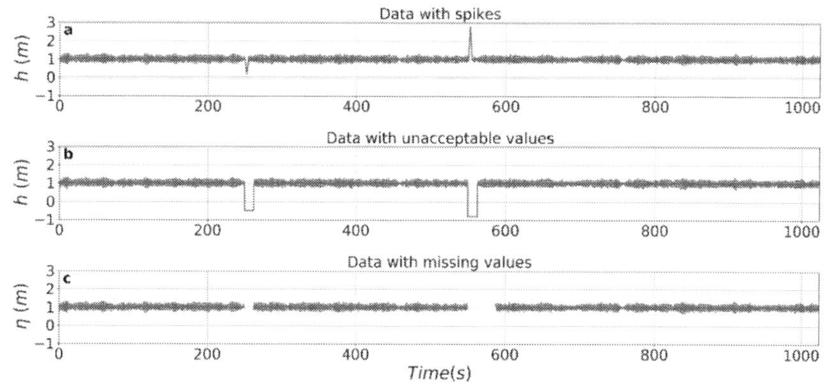

Figure 2.2 Unacceptable data in a time series. Plot (a) shows one burst of data with two spikes at 250 and 550 seconds. Plot (b) shows one burst of data with unacceptable values (negative depth) around 250 and 550 seconds. Plot (c) shows one burst of data with missing values around 250 and 550 seconds. In a plot (a), two spikes can be replaced from adjacent values. In plots (b) and (c), due to a large number of affected points, unacceptable and missing values need to be replaced from another source, if obtainable. Otherwise, unacceptable and missing values should be removed from a time series. Then, a remaining of the dataset should be evaluated to see if it is still acceptable for analysis or not. A time series shows one burst of data with $t_{br} = 1024\ s$.

dataset. In a case that there is a trend in a burst, a trend needs to be removed from a dataset. In other words, a dataset in each burst should be de-trended in a way that a mean value of the signal time series becomes zero throughout each burst. A presence of the trend in a time series might have different reasons. For example, a trend in data might be caused by a measurement method.

For example, assume water surface oscillations, i.e. surface waves, are intended to be measured by deploying a pressure sensor at a seafloor. Measurements by a pressure sensor are results of the pressure from water depth plus pressure from wave oscillations. To obtain water depth values, a time series of the measured pressure data is converted to a time series of the equivalent water depths. A resultant water depth values in this time series are equal to a mean water depth plus waves. Now, to get a pure water surface oscillation time series, a mean value of the water depth needs to be removed from a time series. To do that, at first, a time series is split into a series of

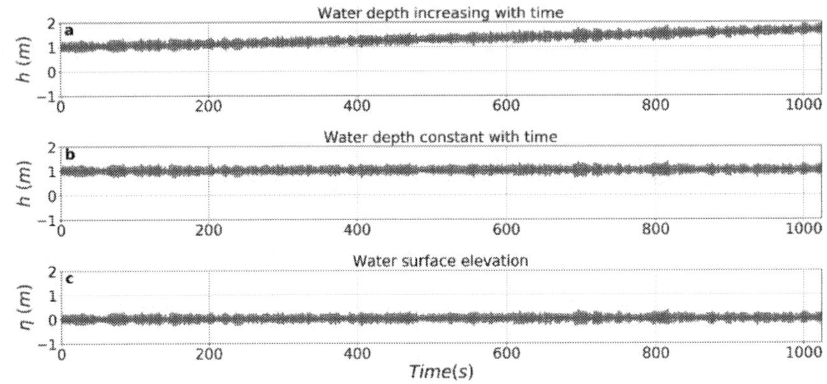

Figure 2.3 De-trending a time series. Plot (a) shows non-stationary water depth, as depth changes with time. Plot (b) shows stationary water depth, as depth is constant with time. Plot (c) shows de-trended water surface elevation. In a plot (a), there is an increasing trend in data, therefore these data cannot be used for signal analysis. In a plot (b), there is a constant non-zero trend (value) in the signal. Data in a plot (c) are de-trended values of the data in a plot (b), resulting in a constant mean value equal to zero throughout a burst. A time series shows one burst of data with $t_{br} = 1024\ s$.

bursts (windows with a definite duration). Then, a mean depth of each burst is subtracted from values in that burst to obtain wave oscillations in that burst, given that data in the burst are stationary. The result of this procedure is the de-trended water surface oscillations caused by wave activities.

If a trend in data is changing by time, it means that data points in a burst are not stationary. In that case, a shorter period of time, Δt_{bl}, should be selected for analysis. For example, assume a time series of water depth data is collected by a pressure sensor located at a seafloor. Then this time series is split into a series of two-hour bursts, i.e. $\Delta t_{bl} = 2\ hr$. If a mean of the water depth is changing throughout any burst, then data in that burst are not stationary. In that case, a time series should be split into a shorter data block such as $\Delta t_{bl} = 1\ hr$ or $\Delta t_{bl} = 30\ min$ to make data within each burst stationary.

3 Introduction to Time Domain Data Analysis: Zero-Crossing Method

In this chapter, random sea and zero-crossing method are covered. First, the random sea is explained. Next, the zero-crossing method is discussed, and then wave properties obtained from the zero-crossing method is explained.

3.1 Random sea

Ocean waves can be linear or nonlinear, regular or irregular, and/or random. Linear and nonlinear waves are explained previously in chapter 1. Except for a rare condition, such as swell waves, water waves are irregular and random in nature. Regular waves can be presented with a single wave frequency (wave period) and a single wave amplitude (wave height). However, irregular and random waves consist of a series of independent frequencies and amplitudes. To represent these random waves, they are considered as a superposition of a series of regular waves as:

$$\eta = \sum a_i \cos(k_i x - \omega_i t + \phi_i) \tag{3.1}$$

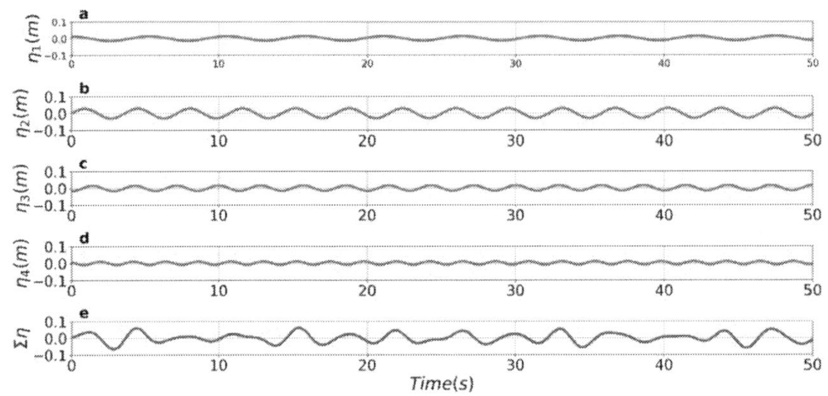

Figure 3.1 Superposition of 4 regular waves to generate an irregular wave time series at $x = 0$. Waves in plots (a) to (d) have independent wave frequencies (wave periods) and wave amplitudes (wave heights). Waves in a plot (e) are resulted from a superposition of the plots (a) to (d), i.e. $\sum \eta = \eta_1 + \eta_2 + \eta_3 + \eta_4$. Wave amplitudes, periods, and phases for plots (a) to (b) are $a = 0.013, 0.029, 0.016, 0.009\ m$, $T = 5.27, 3.58, 2.86, 2.22\ s$, and $\phi = 0, 0.5\pi, \pi, 1.75\pi\ radian$, respectively.

Typically, waves are measured at a stationary location. Therefore, by considering a position of the measurement to be located at $x = 0$, then a time series of the random waves is written as:

$$\eta(t) = \sum a_i \cos(-\omega_i t + \phi_i) \tag{3.2}$$

where, η is a water surface elevation, a is an amplitude, ω is an angular frequency, k is a wavenumber, x is a distance, t is time, $(kx - \omega t)$ is a wave phase in $radian$, the ϕ is a random phase angle or a phase shift in $radian$. Following this view, irregular and random waves are analyzed by decomposing them into a series of monochromic regular waves. This process can be done either in the time domain by using the zero-crossing method or in the frequency domain by using the spectral analysis.

3.2 Zero-crossing method

It was explained previously that a time series of data needs to be quality controlled, and then be split into a series data bursts (windows of data with

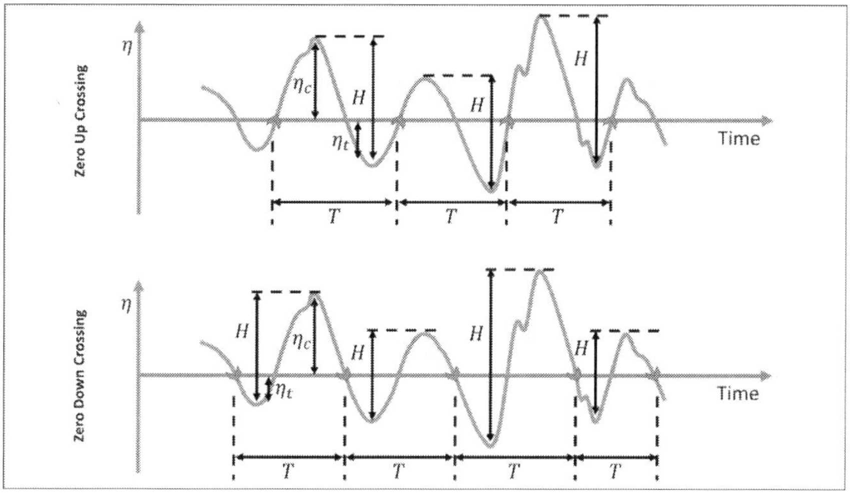

Figure 3.2 The zero-crossing method. Top shows a zero up crossing method and bottom shows a zero down crossing method. *H* and *T* are wave height and period obtained from the zero-crossing method, respectively. η_c and η_t (both are positive) are a wave crest height and a wave trough height, respectively. A crest height is a height of the highest point on a wave from a mean water level. A trough height is a distance from a mean water level to the lowest point on a wave. η_c and η_t are only shown for the first wave.

a definitive duration) before it can be analyzed. After wave data are evaluated, split into a series of bursts, de-trended and verified to be stationary, then they become ready for wave analysis. One method to explore a wave data is to analyze a time series in the time domain. For that purpose, wave data in each de-trended burst are separated into a series of single waves. To separate waves from each other, each wave is defined by two successive points, where at both points, a water surface elevation crosses a zero line in the same direction. It means that, each wave is defined between two successive points where both are crossing a horizontal zero level upward, or both are crossing a horizontal zero level downward. If a wave is defined by two successive points where a water level is crossing up a horizontal axis, it is called an upward zero-crossing or a zero up crossing. However, if a wave is defined by two successive points where a water level is crossing down a horizontal axis, it is called a downward zero-crossing or a zero down crossing.

3.3 Wave properties from zero-crossing method

To start zero-crossing method, first, wave data in each burst are separated into a series of single waves. Then, each burst of data is analyzed separately. To do that, all waves in a wave record (for one burst) are sorted from the largest one to the smallest one. Then, wave properties for that burst are defined as:

$$H_z = \frac{1}{N}\sum_{i=1}^{N} H_i \tag{3.3}$$

$$T_z = \frac{1}{N}\sum_{i=1}^{N} T_i \tag{3.4}$$

where, N is a total number of the waves in the wave record or a burst, sorted from the largest to the smallest. H is wave height, and T is a wave period for each wave defined by the zero-crossing method. H_z is a zero-crossing mean wave height, and T_z is a zero-crossing mean wave period.

In addition to the mean wave properties of the wave record, there are other common values that researchers mostly prefer to use them for wave studies. One is the root mean square (rms) wave height, H_{rms}. Later, it is explained that H_{rms} value is used as a substitute of H value in the linear wave equations. The other parameter is the significant wave height, H_s, which is widely used by researchers to characterize a wave field. H_s is a mean value of the highest one-third of the wave heights. In other words, if wave heights are sorted from the largest to the smallest one, H_s is a mean value of the first 33.33% wave heights. Similarly, if wave periods are sorted from the largest to the smallest one, then a significant wave period, T_s, is defined as a mean value of the highest one-third of the wave periods. If a total of N waves are defined by the zero-crossing method (i.e. if there are N waves in the wave record or a burst), and waves are sorted from the largest to the smallest one, then these parameters are calculated as:

$$H_{rms} = \left(\frac{1}{N}\sum_{i=1}^{N} H_i^2\right)^{0.5} \tag{3.5}$$

$$H_s = \frac{1}{N/3}\sum_{i=1}^{N/3} H_i \tag{3.6}$$

$$T_s = \frac{1}{N/3}\sum_{i=1}^{N/3} T_i \tag{3.7}$$

where, H_{rms} is the root mean square wave height, H_s is significant wave height, and T_s is a significant wave period. It is accepted that if wave heights are narrow banded, i.e. if wave periods have a small range, then wave heights can be well approximated by the Rayleigh distribution. In this case, a probability density function of the wave height, $f_{pdf}(H)$, mean wave height, $\bar{H}$, and root mean square wave height, H_{rms}, following the Rayleigh distribution are written as (Longuet-Higgins, 1952; Young, 1999):

$$f_{pdf}(H) = \left(\frac{H}{4\sigma^2}\right)\exp\left(-\frac{H^2}{8\sigma^2}\right) \tag{3.8}$$

$$\bar{H} = \int_0^\infty H f_{pdf}(H)dH = \sqrt{2\pi\sigma^2} \tag{3.9}$$

$$H_{rms}^2 = \int_0^\infty H^2 f_{pdf}(H)dH = 8\sigma^2 \tag{3.10}$$

where, $f_{pdf}(H)$ is a probability density function of (H), the H is wave height, $\bar{H}$ is mean wave height, H_{rms} is root mean square wave height, and $\sigma = \sqrt{8}H_{rms}$ is a variance of the data. Following the Rayleigh distribution, a probability of the wave height being equal or larger than given wave height, i.e. $P(H \geq H_e)$, is written as (e.g. Dean and Dalrymple, 1991; Holthuijsen, 2007; McCormick, 2009):

$$P(H \geq H_e) = \exp\left(-\left(\frac{H_e}{H_{rms}}\right)^2\right) = \exp\left(-2\left(\frac{H_e}{H_s}\right)^2\right) \tag{3.11}$$

where, $P(H \geq H_e)$ is a probability of $(H \geq H_e)$, the H is wave height, and H_e is a given wave height. Similarly, a probability of the wave height being equal or smaller than a given wave height, i.e. $P(H \leq H_e)$, is:

$$P(H \leq H_e) = 1 - \exp\left(-\left(\frac{H_e}{H_{rms}}\right)^2\right) = 1 - \exp\left(-2\left(\frac{H_e}{H_s}\right)^2\right) \tag{3.12}$$

Using a previous relationship and following the Rayleigh distribution, a probability density function of the wave height, $f_{pdf}(H)$, can be derived as:

$$f_{pdf}(H) = \frac{d}{dH}\left(P(H \leq H_e)\right) = \left(\frac{2H}{H_{rms}^2}\right)\exp\left(-\left(\frac{H}{H_{rms}}\right)^2\right) \tag{3.13}$$

Note that, the total area under a probability density function is unity, i.e. $\int_0^\infty f_{pdf}(H)\, dH = 1$.

Again, consider a wave record consists of N waves sorted from the largest one to the smallest one. Assume $H_{e(n/N)}$ is a given wave height which is ranked n in a wave record, where $n = 1$ represents the largest wave and $n = N$ represents the smallest wave in a wave record. In this case, a total number of waves that are equal or larger than this given wave height, i.e. $H_{e(n/N)}$, are equal to n. This helps to write a probability of H being equal or larger than $H_{e(n/N)}$, i.e. $P(H \geq H_{e(n/N)})$, as:

$$P(H \geq H_{e(n/N)}) = \exp\left(-\left(\frac{H_{e(n/N)}}{H_{rms}}\right)^2\right) = \frac{n}{N} \tag{3.14}$$

where, $P(H \geq H_{e(n/N)})$ is a probability of $(H \geq H_{e(n/N)})$, the H is wave height, and $H_{e(n/N)}$ is a given wave height that is ranked n in a descending sorted wave record. Then $H_{e(n/N)}$ is:

$$H_{e(n/N)} = H_{rms}\sqrt{\ln\left(\frac{N}{n}\right)} = \frac{H_S}{\sqrt{2}}\sqrt{\ln\left(\frac{N}{n}\right)} \tag{3.15}$$

Then, the largest wave height in a wave record, H_{max}, is defined when $n = 1$ as:

$$H_{max} = H_{e(1/N)} = H_{rms}\sqrt{\ln\left(\frac{N}{1}\right)} = \frac{H_S}{\sqrt{2}}\sqrt{\ln\left(\frac{N}{1}\right)} \tag{3.16}$$

For example, consider wave data that are recorded for 1024 s, and have a mean wave period $T_z = 4\ s$. Using a mean wave period, a total number of waves in this record can be approximated as $N = 1024/4 = 256$ waves. Then, a maximum wave height in this record is equal to $H_{max} = H_{rms}\sqrt{\ln(N)} = H_{rms}\sqrt{\ln(256)}$ or $H_{max} \approx 2.355 H_{rms}$.

Furthermore, an average wave height of the top n waves, $H_{n/N}$, i.e. a mean value of the highest n wave heights in a wave record, is (e.g. Dean and Dalrymple, 1991):

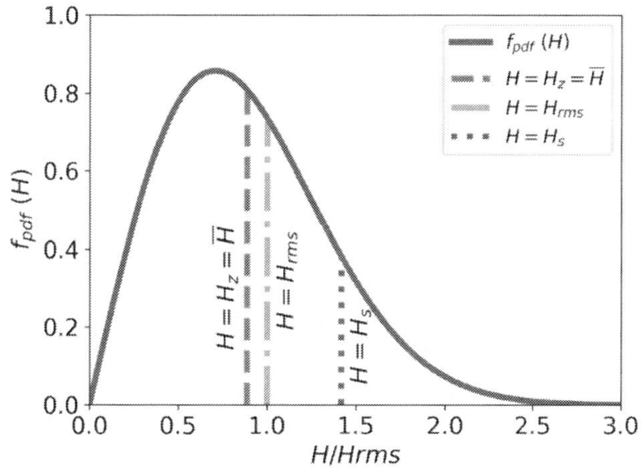

Figure 3.3 Wave height Rayleigh distribution, $f_{pdf}(H)$, for $0 \leq H \leq 3\ m$ and $H_{rms} = 1\ m$. The first vertical line from the left represents $H = H_z = \overline{H}$ or $H/H_{rms} = H_z/H_{rms} \approx 0.866$. The second vertical line from the left represents $H = H_{rms}$ or $H/H_{rms} = H_{rms}/H_{rms} = 1$. The third vertical line from the left represents $H = H_s$ or $H/H_{rms} = H_s/H_{rms} \approx \sqrt{2}$.

$$H_{n/N} = \frac{\int_{H_{e(n/N)}}^{\infty} H f_{pdf}(H) dH}{\int_{H_{e(n/N)}}^{\infty} f_{pdf}(H) dH} \tag{3.17}$$
$$= \left(\sqrt{\ln\left(\frac{N}{n}\right)} + \frac{\sqrt{\pi}}{2(n/N)} erfc\left(\sqrt{\ln\left(\frac{N}{n}\right)}\right)\right) H_{rms}$$

where, $H_{n/N}$ is a mean wave height of the top n waves in a descending sorted wave record. For a mean wave height, $n = N$ and $n/N = 1$, and for significant wave height, $n = N/3$ and $n/N = 1/3$. By using the $H_{n/N}$ equation, a series of important relationships between different wave height definitions are obtained as:

$$H_z = \overline{H} = H_1 = \frac{\int_0^{\infty} H f_{pdf}(H) dH}{\int_0^{\infty} f_{pdf}(H) dH} \tag{3.18}$$
$$= \frac{\int_0^{\infty} H f_{pdf}(H) dH}{1} = \frac{\sqrt{\pi}}{2} H_{rms} \approx 0.866 H_{rms}$$

$$H_s = H_{1/3} = \frac{\int_{H_{e(1/3)}}^{\infty} H f_{pdf}(H) dH}{\int_{H_{e(1/3)}}^{\infty} f_{pdf}(H) dH} \approx \sqrt{2} H_{rms} \tag{3.19}$$

$$H_{1/10} = \frac{\int_{H_{e(1/10)}}^{\infty} H f_{pdf}(H) dH}{\int_{H_{e(1/10)}}^{\infty} f_{pdf}(H) dH} \approx 1.8 H_{rms} \tag{3.20}$$

where, H_z is a zero-crossing mean wave height, H_s is significant wave height, $H_{1/10}$ is an average value of top 10% wave heights, and H_{rms} is the root mean square wave height.

Another relationship between water surface elevation height and wave height is:

$$H_{rms} = 2\sqrt{2}\eta_{rms} = \frac{2}{\sqrt{\pi}} H_z \approx 1.13 H_z \tag{3.21}$$

where, η_{rms} is a root mean square of the water surface elevation.

4 Introduction to frequency domain data analysis: spectral analysis method

In this chapter, random sea analysis using the Fourier and spectral analysis is explained. First, relationships between the time and frequency domains are covered. Next, the Fourier analysis and a power spectral density are discussed. Then, wave properties calculation from a spectral analysis method is explained. Sample codes to analyze data and to calculate wave properties are presented in MATLAB, GNU Octave, and Python.

4.1 Relationships between time and frequency domains

Waves in coastal areas and oceans are typically irregular and random waves. These random waves are formed from a combination of waves with independent frequencies and amplitudes. It is a common practice to consider these random waves as a superposition of a series of regular waves (sinusoidal waves). It means that a random wave in an ocean can be decomposed into a series of sinusoidal cosine waves as:

$$\eta(x,t) = \sum a_i \cos(k_i x - \omega_i t + \phi_i) \qquad (4.1)$$

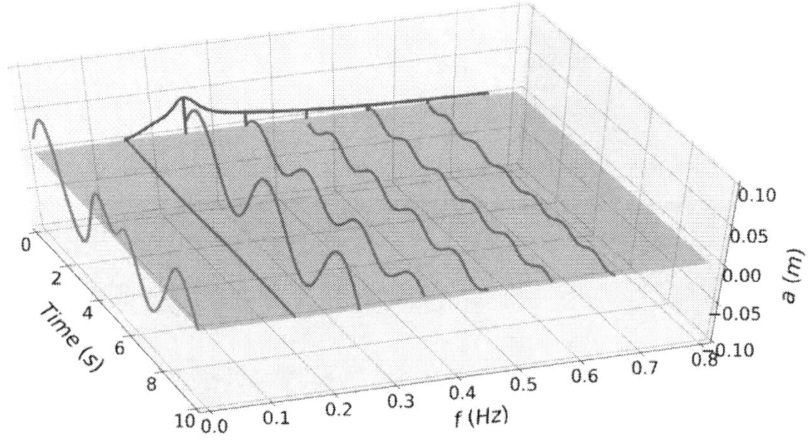

Figure 4.1 Time and frequency domains relationships in a 3D plot. A horizontal x axis is a frequency, f, a horizontal y axis is time, and a vertical z axis is an amplitude, a. First, a random wave time series (a green line) is decomposed into a series of sinusoidal waves (an orange lines). Then, amplitudes of these sinusoidal waves are projected into the frequency-amplitude plot (violet and blue lines). A time series is generated from the JONSWAP spectrum (Hasselmann et al., 1973) for a wind velocity of 10 m/s, a wind fetch of 10000 m, and a peak wave frequency of $f_p = 0.25\ Hz$. A time series has a duration of 1024 s and a sampling frequency of $f_s = 2\ Hz$. Only 10 seconds of the time series is plotted. Decomposed sinusoidal waves are only plotted for a frequency range of $\sim 0.15 - 0.75\ Hz$ in a $0.1\ Hz$ interval.

Now, if a wave time series is measured at a stationary position located at $x = 0$, a wave time series becomes:

$$\eta(t) = \sum a_i \cos(\omega_i t + \phi_i) = \sum a_i \cos\left(\frac{2\pi}{T_i} t + \phi_i\right) = \sum a_i \cos(2\pi f_i t + \phi_i) \tag{4.2}$$

where, η is a water surface elevation, a is an amplitude, $\omega = 2\pi/T = 2\pi f$ is an angular frequency, k is a wavenumber, x is a distance, t is time, $(kx - \omega t)$ is a wave phase in *radian*, the ϕ is a random phase angle or a phase shift in *radian*, the $T = 1/f$ is a wave period, and f is a wave frequency.

Presenting a random wave as a combination of sinusoidal waves allows moving data from the time domain to the frequency domain. To explain how to do that, assume a random wave time series is decomposed into a series of sinusoidal wave time series. Each of these sinusoidal waves can be presented by an ordered pair of the frequency and amplitude as (f_i, a_i). Now, wave data are presented in the frequency domain by plotting each pair of (f, a) in a frequency-amplitude plot. In fact, by doing this, wave data from the time domain are transformed into the frequency domain.

4.2 Fourier analysis

To start wave analysis in the frequency domain, first, wave data need to be transformed from the time domain to the frequency domain. To transform data from time to frequency domain, data should be decomposed into a series of waves with known amplitudes and frequencies. Before explaining a decomposition process, first, it is demonstrated how to write a periodic function as a sum of simple waves (sines and cosines). In theory, the Fourier series can be used for this purpose. By using the Fourier series, a time series of the surface elevation is written as a superposition of its components as (e.g. Chakrabarti, 1987; Holthuijsen, 2007; O'neil, 2011; Reeve et al., 2012; Dyke, 2014):

$$\eta(t) = \frac{a_0}{2} + \sum_{n=1}^{N} (a_n \cos(n\omega t) + b_n \sin(n\omega t)) \tag{4.3}$$

$$a_0 = \frac{2}{T} \int_{-T/2}^{T/2} \eta(t) dt \tag{4.4}$$

$$a_n = \frac{2}{T} \int_{-T/2}^{T/2} \eta(t) \cos(n\omega t)\, dt \tag{4.5}$$

$$b_n = \frac{2}{T} \int_{-T/2}^{T/2} \eta(t) \sin(n\omega t)\, dt \tag{4.6}$$

The Fourier series can be written in a discrete complex form as:

$$\eta(t) = \sum_{n=-N}^{N} c_n e^{in\omega t} \tag{4.7}$$

$$c_n = \frac{1}{T} \int_{-T/2}^{T/2} \eta(t) e^{-in\omega t} dt = \frac{a_n - ib_n}{2} \tag{4.8}$$

$$c_{-n} = \frac{1}{T} \int_{-T/2}^{T/2} \eta(t) e^{in\omega t} dt = \frac{a_n + ib_n}{2} \tag{4.9}$$

$$c_0 = \frac{a_0}{2} \tag{4.10}$$

$$|c_n| = \frac{1}{2}\sqrt{a_n^2 + b_n^2} \tag{4.11}$$

Also, it can be written in a phase angle form as:

$$\eta(t) = \frac{a_0}{2} + \sum_{n=1}^{N} d_n \cos(n\omega t + \emptyset_n) \tag{4.12}$$

$$d_n = \sqrt{a_n^2 + b_n^2} = 2|c_n| \tag{4.13}$$

$$\emptyset_n = \tan^{-1}\left(-\frac{b_n}{a_n}\right) \tag{4.14}$$

where, $\eta(t)$ is a time series of the water surface elevation, a_n, b_n, c_n and d_n are amplitudes (Fourier series coefficients), $\emptyset_n$ is a phase, $\omega = 2\pi/T = 2\pi f$ is an angular frequency, and $i = \sqrt{-1}$ is an imaginary unit. The c_n is a complex number, where $|c_n|$ shows a magnitude of c_n. Note that $b_0 = 0$. Also, $(a_0/2)$ is equal to an average value of $\eta(t)$ over an interval of $[-T/2, T/2]$. Since an average value of the water surface fluctuation is equal to zero, then $(a_0/2) = 0$ for a water surface time series.

An implementation of the Fourier series shows that a time series can be written into a series of waves with known amplitudes and frequencies. Now, to transform data from the time domain to the frequency domain, these amplitudes and frequencies need to be defined. To define these amplitudes and frequencies, a random wave time series must be decomposed into a series of sinusoidal waves. To achieve this decomposition, the Fourier transform is implemented. The Fourier transform converts an original signal in an equally spaced time domain into a signal in an equally spaced frequency domain by returning frequencies and amplitudes of the waves that form the original signal. The continuous Fourier transform of the surface elevation time series, $\mathcal{F}\{\eta(t)\}(\omega)$ or simply $F(\omega)$, and an inverse Fourier transform, $\mathcal{F}^{-1}\{F(\omega)\}(t)$ or simply $\eta(t)$, are written as:

$$F(\omega) = \mathcal{F}\{\eta(t)\}(\omega) = \int_{-\infty}^{\infty} \eta(t) e^{-i\omega t}\, dt \tag{4.15}$$

$$\eta(t) = \mathcal{F}^{-1}\{F(\omega)\}(t) = \frac{1}{2\pi}\int_{-\infty}^{\infty} F(\omega)e^{i\omega t}d\omega \tag{4.16}$$

where, $\mathcal{F}$ is a Fourier transform operator and $\mathcal{F}^{-1}$ is an inverse Fourier transform operator.

In the term of the frequency, f, a continuous Fourier transform, $F(f)$, and its inverse, $\eta(t)$, are:

$$F(f) = \int_{-\infty}^{\infty} \eta(t)e^{-i2\pi ft}dt \tag{4.17}$$

$$\eta(t) = \int_{-\infty}^{\infty} F(f)e^{i2\pi ft}df \tag{4.18}$$

A discrete Fourier transform, $\mathcal{F}\{\eta[t_k]\}[\omega_n]$ or simply $F[\omega_n]$, and its inverse, $\mathcal{F}^{-1}\{F[\omega_n]\}[t_k]$ or simply $\eta[t_k]$, are (Smith, 2007; Shima and Nakayama; 2010):

$$F[\omega_n] = \mathcal{F}\{\eta[t_k]\}[\omega_n] = \sum_{k=0}^{N-1} \eta[t_k]e^{-i\omega_n t_k} \tag{4.19}$$

$$\eta[t_k] = \mathcal{F}^{-1}\{F[\omega_n]\}[t_k] = \frac{1}{N}\sum_{n=0}^{N-1} F[\omega_n]e^{i\omega_n t_k} \tag{4.20}$$

By considering $\omega_n = n\omega = n(2\pi/T)$, $T = N\Delta t$, and $t_k = k\Delta t$, a discrete Fourier transform, $F[n]$, and its inverse, $\eta[k]$, are (Smith, 2007; Shima and Nakayama; 2010):

$$F[n] = \sum_{k=0}^{N-1} \eta[k]e^{-i2\pi n\frac{k}{N}} \tag{4.21}$$

$$\eta[k] = \frac{1}{N}\sum_{n=0}^{N-1} F[n]e^{i2\pi k\frac{n}{N}} \tag{4.22}$$

By considering $T = N\Delta t$, $t_k = k\Delta t$, and $f_n = 1/T_n = n/T = n/(N\Delta t)$, and using an absolute symbol, $|\ |$, to show a magnitude of the complex number, coefficients of the Fourier series are calculated from a discrete Fourier transform as (Smith, 2007; Kawauchi, 2009; Shima and Nakayama; 2010):

$$c_n = \frac{1}{T}F\left(\frac{n}{T}\right) = \frac{1}{T}\int_{-T/2}^{T/2} \eta(t)e^{-i2\pi\frac{n}{T}t}dt \approx \frac{1}{N}\sum_{n=0}^{N-1} \eta[k]e^{-i2\pi n\frac{k}{N}} \approx \frac{F[n]}{N} \tag{4.23}$$

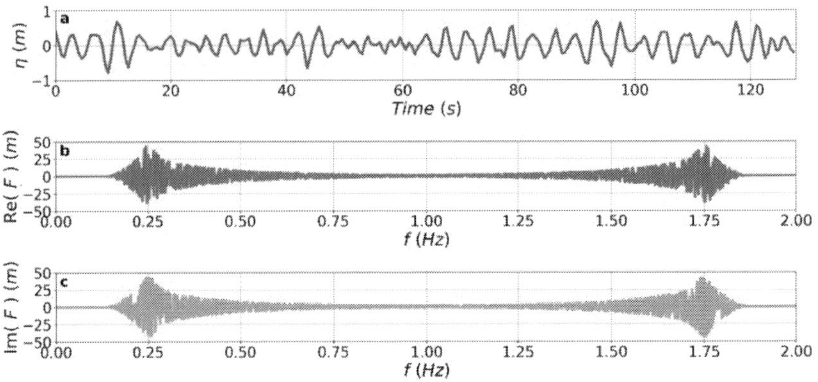

Figure 4.2 Time domain to frequency domain transformation using the Fourier transform. The Fourier transform is presented in a complex form of $F = Re(F) + iIm(F)$. Plot (a) shows a time series of the water surface elevation in the time domain. Plot (b) represents a real part of the Fourier transform, i.e. $Re(F)$. Plot (c) represents an imaginary part of the Fourier transform, i.e. $Im(F)$. A time series in a plot (a) is generated from the JONSWAP spectrum (Hasselmann et al., 1973) for a wind velocity of 10 m/s, a wind fetch of 10000 m, and a peak wave frequency of 0.25 Hz. A time series has a duration of 1024 s and a sampling frequency of $f_s = 2\ Hz$. Only 128 seconds of the time series is plotted.

$$d_n = 2|c_n| = 2\left|\frac{1}{T}F\left(\frac{n}{T}\right)\right| = 2\left|\frac{F[n]}{N}\right| \tag{4.24}$$

$$F(f) \approx \sum_{k=0}^{N-1}\eta(t_k)e^{-i2\pi f_n t_k}\Delta t \approx \sum_{k=0}^{N-1}\eta(t_k)e^{-i2\pi\left(\frac{n}{N\Delta t}\right)(k\Delta t)}\Delta t \approx \Delta t\sum_{k=0}^{N-1}\eta[k]e^{-i2\pi n\frac{k}{N}} \approx \Delta t \times F[n] \tag{4.25}$$

A summary of the relationships between period and frequency definitions are:

$$T = N\Delta t \tag{4.26}$$

$$T_n = \frac{T}{n} = \frac{N\Delta t}{n} \tag{4.27}$$

$$f_s = \frac{1}{\Delta t} \tag{4.28}$$

$$\Delta f = f_{n+1} - f_n = \frac{f_s}{N} = \frac{1}{N\Delta t} = \frac{1}{T} \tag{4.29}$$

$$f_n = \frac{n}{T} = \frac{n}{N\Delta t} = n\Delta f \tag{4.30}$$

Eventually, the Fourier series is written as a function of the discrete Fourier transform as:

$$\eta(t) = \sum_{n=-N}^{N} c_n e^{in\omega t} = \sum_{n=-N}^{N} \frac{F[n]}{N} e^{i2\pi\frac{n}{T}t} \tag{4.31}$$

4.3 Power spectral density

One of the main tools to analyze data in the frequency domain is a power spectral density. A power spectral density of the time series describes that how wave powers are distributed in the frequency domain. It is explained that by using the Fourier transform, a time series can be decomposed into a series of sinusoidal waves. After obtaining these sinusoidal waves, a power associated with each wave is calculated. A power spectral density is attained by plotting these powers respect to frequencies associated with each power. Here, it is explained in details how these steps are taken.

One method to calculate a power spectral density is to use the Fourier series in a phase angle form. Consider the Fourier series of the water surface elevation time series as:

$$\eta(t) = \sum_{n=1}^{N} d_n \cos(n\omega t + \emptyset_n) \tag{4.32}$$

or

$$\eta(t) = \sum_{n=1}^{N} d_n \cos(2\pi f_n t + \emptyset_n) \tag{4.33}$$

where, $f_n = n/(N\Delta t)$ is a frequency, and $(N\Delta t)$ is a total duration of the time series. If amplitude for each frequency is known in the form of the ordered pair (f_n, d_n), then wave energy spectrum is defined as:

$$E(f_n) = \frac{1}{8}\rho g H_n^2 = \frac{1}{2}\rho g d_n^2 \tag{4.34}$$

where, E is a wave energy density per unit horizontal area of the water surface with a unit of joule per unit area, J/m^2, the ρ is water density, g is the gravitational acceleration, and $H_n = 2d_n$ is wave height. Now, a water surface elevation power spectral density is calculated as:

$$S_{\eta\eta}(f_n) = \frac{1}{\rho g \Delta f} E(f_n) = \frac{d_n^2}{2\Delta f} \tag{4.35}$$

A water surface elevation power spectral density is calculated in a limit form as:

$$S_{\eta\eta}(f_n) = \frac{1}{\Delta f} \times \sum_f^{f+\Delta f} \left(\frac{1}{2} d_n^2\right) = \lim_{\Delta f \to 0} \frac{1}{\Delta f} \left(\frac{1}{2} d_n^2\right) \tag{4.36}$$

where, $S_{\eta\eta}$ is a one-sided water surface elevation power spectral density with a unit of m^2/Hz or $m^2 s$, which represent an energy density per frequency. $f_n = n/(N\Delta t)$ is a frequency, $\Delta f = f_{n+1} - f_n = f_s/N = 1/(N\Delta t)$ is a frequency interval in Hz, and $(N\Delta t)$ is a total duration of the time series.

In another method, a power spectral density is obtained by substituting the Fourier series in a wave energy equation. To do that, first, a wave energy equation need to be obtained. Wave energy is considered as a sum of the potential and kinetic energies. Mean values of the potential, kinetic and total wave energies are:

$$E_P = E_K = \frac{1}{T} \int_0^T \frac{1}{2} \rho g \eta^2(t) dt \tag{4.37}$$

$$E = E_P + E_K = \frac{1}{T} \int_0^T \rho g \eta^2(t) dt \tag{4.38}$$

where, E_P, E_K, and E are mean values of the potential, kinetic and total energy densities with a unit of joule per unit area, J/m^2, respectively. Now, by substituting a $\eta(t)$ with the Fourier series of the water surface elevation, a total energy equation is rewritten as (Parseval's theorem):

$$E = \frac{1}{T} \int_0^T \rho g \left(\frac{a_0}{2} + \sum_{n=1}^{N} (a_n \cos(n\omega t) + b_n \sin(n\omega t))\right)^2 dt \tag{4.39}$$

or

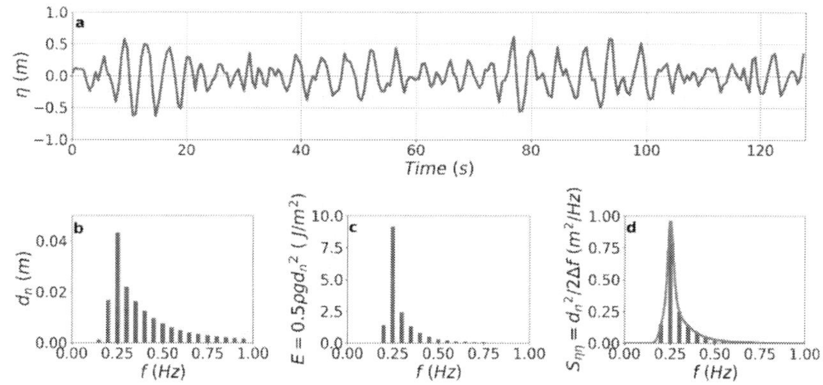

Figure 4.3 Demonstration of the time domain to a power spectral density transformation. Plot (a) shows a time series of the water surface elevation in the time domain. Plot (b) shows a wave amplitude, d_n, in m. Plot (c) shows energy associated with each frequency as $E = 0.5\rho g d_n^2$ in J/m^2. Plot (d) shows a surface elevation power spectral density as $S_{\eta\eta} = d_n^2/2\Delta f$ in m^2/Hz. A time series in a plot (a) is generated from the JONSWAP spectrum (Hasselmann et al., 1973) for a wind velocity of 10 m/s, a wind fetch of 10000 m, and a peak wave frequency of 0.25 Hz. A time series has a duration of 1024 s and sampling frequency of $f_s = 2\ Hz$. Only 128 seconds of the time series is plotted.

$$E = \frac{1}{T}\int_0^T \rho g \left(\frac{a_0}{2} + \sum_{n=1}^{N}(a_n \cos(2\pi f_n t) + b_n \sin(2\pi f_n t))\right)^2 dt \tag{4.40}$$

As $a_0/2$ represents a mean value of the water surface elevation, then $a_0/2 = 0$, and therefore, the total energy is:

$$E = \rho g \sum_{n=1}^{N}\left(\frac{a_n^2+b_n^2}{2}\right) = \rho g \sum_{n=1}^{N}(2|c_n|^2) = \rho g \sum_{n=1}^{N}\left(\frac{1}{2}d_n^2\right) \tag{4.41}$$

$$E(f_n) = \sum_f^{f+\Delta f}(E) \approx \frac{1}{2}\rho g d_n^2 \tag{4.42}$$

Then, a water surface elevation power spectral density is:

$$S_{\eta\eta}(f_n) = \frac{1}{\rho g \Delta f}E(f_n) = \frac{d_n^2}{2\Delta f} \tag{4.43}$$

A water surface elevation power spectral density in a limit form is:

$$S_{\eta\eta}(f_n) = \frac{1}{\rho g \Delta f} \times \sum_f^{f+\Delta f}(E) = \lim_{\Delta f \to 0} \frac{1}{\Delta f}\left(\frac{1}{2} d_n^2\right) \approx \frac{d_n^2}{2\Delta f} \tag{4.44}$$

In these two methods, it is demonstrated how to calculate a power spectral density from the values of (f_n, c_n) or (f_n, d_n). However, in practice, these two methods are more favorable to back calculate (f_n, c_n) or (f_n, d_n) from an available power spectral density. In other words, if a power spectral density is known, then one of these two methods is used to calculate (f_n, c_n) or (f_n, d_n). Knowing values of (f_n, c_n) or (f_n, d_n) helps to reconstruct a time series from a spectrum.

4.4 Periodogram method

A common method to calculate a power spectral density is to use a Fourier transform and periodogram method. The Fourier transform of water surface elevation is:

$$F(f) = \int_{-\infty}^{\infty} \eta(t) e^{-i2\pi f t} dt \tag{4.45}$$

By considering $T = N\Delta t$, $t_k = k\Delta t$, and $f_n = 1/T_n = n/T = n/(N\Delta t)$, the Fourier transform is estimated as (Earle, 1996; Smith, 2007; Kawauchi, 2009):

$$F(f) \approx \sum_{k=0}^{N-1} \eta(t_k) e^{-i2\pi f_n t_k} \Delta t \approx \sum_{k=0}^{N-1} \eta(t_k) e^{-i2\pi\left(\frac{n}{N\Delta t}\right)(k\Delta t)} \Delta t \\ \approx \Delta t \sum_{k=0}^{N-1} \eta[k] e^{-i2\pi n \frac{k}{N}} \approx \Delta t \times F[n] \tag{4.46}$$

Then, a two-sided power spectral density is:

$$S_{\eta\eta-2S}(f_n) = \frac{1}{N\Delta t} \times |F(f)|^2 = \frac{1}{N\Delta t} \times |\Delta t \times F[n]|^2 \\ = \frac{(\Delta t)^2}{N\Delta t} \times |F[n]|^2 \tag{4.47}$$

In the term of the sampling frequency, $f_s = \frac{1}{\Delta t}$, a two-sided power spectral density is:

$$S_{\eta\eta-2S}(f_n) = \frac{1}{N\Delta t f_s^2} \times |F[n]|^2 = \frac{1}{N f_s} \times |F[n]|^2 \tag{4.48}$$

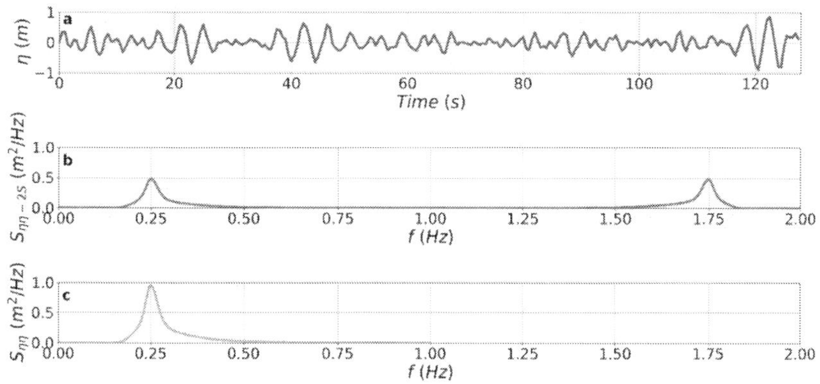

Figure 4.4 Demonstration of the time domain to a two-sided and a one-sided power spectral density transformation. Plot (a) shows a time series of the water surface elevation in the time domain. Plot (b) shows a two-sided power spectral density, $S_{\eta\eta-2S}$, in m^2/Hz with a frequency range of $0 \le f \le f_s - \Delta f$. Plot (c) shows a one-sided power spectral density, $S_{\eta\eta}$, in m^2/Hz with a frequency range of $0 \le f \le f_s/2$. A time series and spectra represent the JONSWAP spectrum (Hasselmann et al., 1973) for a wind velocity of 10 m/s, a wind fetch of 10000 m, and a peak wave frequency of 0.25 Hz. A time series has a duration of 1024 s and a sampling frequency of $f_s = 2\ Hz$. Only 128 seconds of the time series is plotted.

where, $S_{\eta\eta-2S}(f_n)$ is a two-sided power spectral density with a frequency range of $-f_s/2 \le f \le f_s/2$, i.e. $-f_{Ny} \le f \le f_{Ny}$. The $f_s = 1/\Delta t$ is a sampling frequency, the $f_{Ny} = f_s/2 = 1/(2\Delta t)$ is the Nyquist frequency, Δt is a time interval, N is a total number of samples in a time series, and $N\Delta t$ is a total duration of the measurement period, i.e. a total duration of the time series. Note that $F(f_n)$ is a continuous Fourier transform and $F[n]$ is a discrete Fourier transform. The $F[n]$ is a complex number where $|F[n]|$ shows a magnitude of $F[n]$.

An outcome of the Fourier Transform is symmetrical around $f_s/2$. Therefore, a two-sided power spectral density is symmetrical around $f_s/2$. However, it is a common practice to only use values with a frequency range of $0 \le f \le f_s/2$ in a spectrum calculation. This adoption leads to a one-sided spectrum. To compensate for excluded wave energies with a frequency range of $f_s/2 < f \le f_s$, wave energies with a frequency range of $\Delta f \le f_n \le f_s/2 - \Delta f$ are multiplied by 2. Then, a one-sided power spectral density, $S_{\eta\eta}(f_n)$, is calculated from a two-sided power spectral density as:

$$S_{\eta\eta}(0) = S_{\eta\eta-2S}(0) \tag{4.49}$$

$$S_{\eta\eta}(\Delta f \leq f_n \leq f_s/2 - \Delta f) = 2 \times S_{\eta\eta-2S}(\Delta f \leq f_n \leq f_s/2 - \Delta f) \tag{4.50}$$

$$S_{\eta\eta}(f_s/2) = S_{\eta\eta-2S}(f_s/2) \tag{4.51}$$

where, $S_{\eta\eta}(f_n)$ is a one-sided power spectral density with a frequency range of $0 \leq f \leq f_s/2$ (i.e. $0 \leq f \leq f_{Ny}$), and $\Delta f = f_{n+1} - f_n = f_s/N = 1/(N\Delta t)$ is a frequency interval.

4.5 Frequency ordering

Mathematically, outcomes of the Fourier transform have N frequencies which are spread symmetrically around $f = 0$. These Fourier transform outcomes are equally spaced in the frequency domain with a frequency range of $-f_s/2 + \Delta f \leq f \leq f_s/2$ and a frequency interval of $\Delta f = f_{n+1} - f_n = f_s/N = 1/(N\Delta t)$, where f_s is a sampling frequency, $f_n = n/(N\Delta t)$ is a frequency, Δt is a time interval, N is a total number of samples in a time series, and $(N\Delta t)$ is a total duration of the measurement period, i.e. a total duration of the time series. In a two-sided power spectral density, frequencies are ordered in an equally spaced frequency of Δf as (Mobley, 2014):

$$\left\{\left(-\frac{N}{2}+1\right)\Delta f, \left(-\frac{N}{2}+2\right)\Delta f, \cdots, -\Delta f, 0, \Delta f, \cdots, \left(\frac{N}{2}-1\right)\Delta f, \left(\frac{N}{2}\right)\Delta f\right\} \tag{4.52}$$

Similarly, these frequencies are ordered in the term of the sampling frequency as:

$$\left\{\left(-\frac{f_s}{2}+\Delta f\right), \left(-\frac{f_s}{2}+2\Delta f\right), \cdots, -\Delta f, 0, \Delta f, \cdots, \left(\frac{f_s}{2}-\Delta f\right), \left(\frac{f_s}{2}\right)\right\} \tag{4.53}$$

In practice, programming languages are used to decompose a signal into its frequency components. Programming languages mostly use the fast Fourier transform routine, *FFT*, for this purpose. The outcomes of the fast Fourier transform routine are spread symmetrically around $f = f_s/2$ with a frequency range of $0 \leq f \leq f_s$ and frequency interval of Δf. Therefore, a two-sided power spectral density resulted from the fast Fourier transform routine has frequencies in the following orders:

$$\left\{0, \Delta f, \cdots, \left(\frac{N}{2} - 1\right)\Delta f, \left(\frac{N}{2}\right)\Delta f, \left(-\frac{N}{2} + 1\right)\Delta f, \left(-\frac{N}{2} + 2\right)\Delta f, \cdots, -\Delta f\right\} \tag{4.54}$$

Similarly, orders of its frequencies in the term of the sampling frequency are:

$$\left\{0, \Delta f, \cdots, \left(\frac{f_s}{2} - \Delta f\right), \left(\frac{f_s}{2}\right), \left(-\frac{f_s}{2} + \Delta f\right), \left(-\frac{f_s}{2} + 2\Delta f\right), \cdots, -\Delta f\right\} \tag{4.55}$$

In a one-sided power spectral density, only half of the two-sided spectrum with a frequency range of $0 \leq f \leq f_s/2$ is reserved. Therefore, a one-sided power spectral density has $\left(\frac{N}{2} + 1\right)$ frequencies which are ordered in an equally spaced frequency of Δf. As a result, a one-sided power spectral density resulted from the fast Fourier transform routine has frequencies in the following orders:

$$\left\{0, \Delta f, \cdots, \left(\frac{N}{2} - 1\right)\Delta f, \left(\frac{N}{2}\right)\Delta f\right\} \tag{4.56}$$

Similarly, orders of its frequencies in the term of the sampling frequency are:

$$\left\{0, \Delta f, \cdots, \left(\frac{f_s}{2} - \Delta f\right), \left(\frac{f_s}{2}\right)\right\} \tag{4.57}$$

4.6 Power spectral density calculation code

Now, a step by step practical guide is provided on how to calculate a power spectral density from a signal time series using MATLAB or GNU Octave. Then, a Python code is presented.

In practice, researchers work on time series signals that represent various types of situations, particularly the ones that are gathered from real phenomena. For example, a researcher might work on a water wave signal that is obtained from a measurement in the field or in a laboratory. Here, a simple synthetic time series is generated and used for the purpose of illustration in this example. Any other time series can be used as well. To generate this time series, a duration and a sampling frequency of the time series are given as:

```
%Input parameters
duration=1024; %Total duration of the time series,
duration=N*dt (here: duration=2^10)
fs=2; %Sampling frequency, fs=1/dt
```

Next, a time interval, Δt, and a time vector contains time values are calculated as:

```
%Generating a time vector from 0 to (duration-dt), equally
spaced at dt
dt=1/fs; %Time difference between consecutive samples,
dt=1/fs
t(:,1)=[0:dt:duration-dt]; %Time
```

Then, a simple water surface time series is generated using a sinusoidal wave, i.e. $\eta(t) = a\cos(2\pi f t)$, with an amplitude of $a = 0.5\ m$ and a frequency of $f = 1/T = 0.2\ Hz$ as:

```
%Generating a water surface time series with an amplitude of
0.5 m and a frequency of 0.2 Hz (i.e. a period of 5 s)
Eta(:,1)=0.5.*cos(2*pi*0.2*t);
```

After a sinusoidal signal is generated, a random signal (i.e. noise), is added to a sinusoidal signal as:

```
%Adding a random signal to a sinusoidal signal
rnd(:,1)=-0.1+(0.1-(-0.1)).*rand(length(t(:,1)),1);
%Uniformly distributed random signal between -0.1 m and 0.1
m
Eta=Eta+rnd; %Adding a random signal to a sinusoidal signal
```

For the purpose of the spectral analysis, the first step after a time series is obtained, is to remove any trend in data or to detrend the data. In other words, a mean value of the time series should be removed from a dataset in a way that a mean value of the time series becomes zero. This detrending must be done as the first step for almost any signal analysis calculation.

```
%Detrending the data
Eta=detrend(Eta);
```

Now, a frequency vector containing $\left(\frac{N}{2}+1\right)$ frequency values in a range of $0 \leq f \leq f_s/2$, where $f_s/2 = f_{Ny}$, the f_{Ny} is the Nyquist frequency, and N is a total number of samples in a time series, is generated as:

```
%Generating a frequency vector
N=length(Eta(:,1)); %Length of the time series
df=fs/N; %Frequency difference between consecutive samples,
df=fs/N
f(:,1)=[0:df:fs/2]; %Frequency vector from 0 Hz to fNy=fs/2
Hz, equally spaced at df
```

The fast Fourier transform of the water surface elevation is:

```
%Fast Fourier Transform (FFT) of the water surface elevation
from f=0 Hz to f=fs Hz
EtaFFT=fft(Eta);
```

A total number of elements in the fast Fourier transform output is equal to a total number of elements in a time series. In other words, the output of the fast Fourier transform has the same length as a water surface time series, with a frequency range of $0 \leq f \leq f_s - \Delta f$. It means that if a time series has N data points, then the fast Fourier transform result has N data points as well. To calculate a one-sided power spectral density, only values associated with $0 \leq f \leq f_s/2$ needs to be used. Therefore, values of the fast Fourier transform that are in a frequency range of $0 \leq f \leq f_s/2$ are retained as:

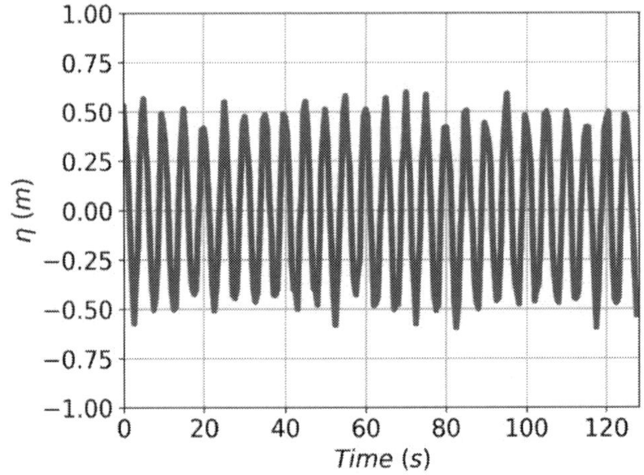

Figure 4.5 Demonstration of the generated water surface elevation time series. A time series is generated from a sinusoidal signal, i.e. $\eta(t) = a\cos(2\pi ft)$, with an amplitude of $a = 0.5\ m$ and a frequency of $f = 1/T = 0.2\ Hz$. A signal is contaminated with a randomly generated noise with a range of $-0.1\ m$ and $0.1\ m$. A time series has a duration of $1024\ s$ and a sampling frequency of $f_s = 2\ Hz$. Only 128 seconds of the time series is plotted.

```
%Fast Fourier Transform (FFT) of the water surface elevation
from f=0 Hz to fNy=fs/2 Hz
EtaFFT=EtaFFT(1:length(f(:,1))); %(N/2+1) elements if N is
even, and (N+1)/2 elements if N is odd
```

Remember that a two-side spectrum has a frequency range of $0 \le f \le f_s - df$. But, here only the first half of the two-sided spectrum is retained. Now, the first half of the two-sided power spectral density which has a frequency range of $0 \le f \le f_s/2$ is calculated as:

```
%Half of the two-sided power spectral density (psd) from f=0
Hz to fNy=fs/2 Hz
psdEta_2Sided=(1/(N*fs))*(abs(EtaFFT)).^2; %Calculating psd
using fs, in (m^2/Hz)
```

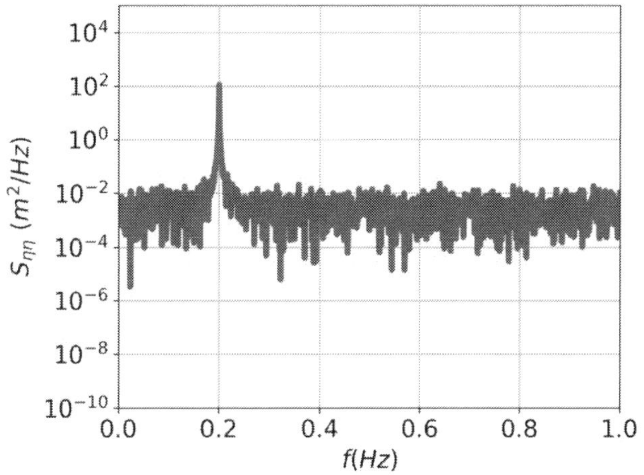

Figure 4.6 Demonstration of the power spectral density obtained from the fast Fourier transform. An analyzed time series has a primary frequency of $f = 0.2\ Hz$ and a sampling frequency of $f_s = 2\ Hz$. A resulted spectrum has a frequency range of $0 \leq f \leq f_s/2$, or $0 \leq f \leq 1$ with a peak wave frequency of $f_p = 0.2\ Hz$.

Values of the one-sided power spectrum are obtained by multiplying the first half of the two-sided spectrum by two, except for $f = 0$ and $f = f_s/2$. It means that $S_{\eta\eta}(0) = S_{\eta\eta-2S}(0)$, $S_{\eta\eta}(\Delta f \leq f_n \leq f_s/2 - \Delta f) = 2 \times S_{\eta\eta-2S}(\Delta f \leq f_n \leq f_s/2 - \Delta f)$, and $S_{\eta\eta}(f_s/2) = S_{\eta\eta-2S}(f_s/2)$. Then, a one-sided power spectral density which has a frequency range of $0 \leq f \leq f_s/2$ is calculated as:

```
%One-sided power spectral density (psd) from f=0 Hz to
fNy=fs/2 Hz
psdEta_1Sided=psdEta_2Sided;
psdEta_1Sided(2:end-1,1)=2.*psdEta_1Sided(2:end-1,1); %one-
sided-spectrum=2*two-sided-spectrum in (m^2/Hz)
```

Next, time series and spectrum data are plotted. A power spectrum is typically illustrated (plotted) in a logarithmic or semi-logarithmic scale. A water surface elevation time series and a power spectral density are plotted as:

```
%Plotting a water surface elevation
figure(1)
plot(t,Eta)
title('Water surface elevation')
xlabel('Time (s)')
ylabel('Surface Elevation (m)')

%Plotting a power spectral density
figure(2)
semilogy(f,psdEta_1Sided)
title('Power spectral density')
xlabel('Frequency (Hz)')
ylabel('Power/Frequency (m^2/Hz)')
```

Here is a Python code for calculating a power spectral density. Note that you need to install and import NumPy package as np, SciPy package as sp, and matplotlib.pyplot package as plt along with Python.

```
#------------------------------------------------------------------
#Importing required packages

import numpy as np
import scipy as sp
from scipy import signal
import matplotlib.pyplot as plt

#------------------------------------------------------------------

#Input parameters
duration=1024; #Total duration of the time series, duration=N*dt
(here: duration=2^10)
fs=2; #Sampling frequency, fs=1/dt

#Generating a time vector from 0 to (duration-dt), equally spaced
at dt
dt=1/fs; #Time difference between consecutive samples, dt=1/fs
t=np.arange(0,duration,dt); #Time

#Generating a water surface time series with an amplitude of 0.5
m and a frequency of 0.2 Hz (i.e. period of 5 s)
Eta=0.5*np.cos(2*np.pi*0.2*t);

#Adding a random signal to a sinusoidal signal
rnd=-0.1+(0.1-(-0.1))*np.random.rand(len(t)); #Uniformly
distributed random signal between -0.1 m and 0.1 m
Eta=Eta+rnd; #Adding random signal to a sinusoidal signal

#Detrending the data
```

```
Eta=sp.signal.detrend(Eta)

#Generating a frequency vector
N=len(Eta); #Length of the time series
df=fs/N; #Frequency difference between consecutive samples,
df=fs/N
f=np.arange(0,fs/2+df,df); #Frequency vector from 0 Hz to
fNy=fs/2 Hz, equally spaced at df

#-----------------------------------------------------------------

#Fast Fourier Transform (FFT) of the water surface elevation from
f=0 Hz to f=fs Hz
EtaFFT=np.fft.fft(Eta);

#Fast Fourier Transform (FFT) of the water surface elevation from
f=0 Hz to fNy=fs/2 Hz
EtaFFT=EtaFFT[0:len(f)]; #(N/2+1) elements if N is even, and
(N+1)/2 elements if N is odd

#Half of the two-sided power spectral density (psd) from f=0 Hz
to fNy=fs/2 Hz
psdEta_2Sided=(1/(N*fs))*(np.absolute(EtaFFT))**2; #Calculating
psd using fs in (m^2/Hz)

#One-sided power spectral density (psd) from f=0 Hz to fNy=fs/2
Hz
psdEta_1Sided=psdEta_2Sided.copy();
psdEta_1Sided[1:-1]=2.*psdEta_1Sided[1:-1]; #one-sided-
spectrum=2*two-sided-spectrum in (m^2/Hz)

#-----------------------------------------------------------------
#Plotting a water surface elevation
plt.figure(1)
plt.plot(t,Eta)
plt.title('Water surface elevation')
plt.xlabel('Time (s)')
plt.ylabel('Surface Elevation (m)')

#-----------------------------------------------------------------
#Plotting a power spectral density
plt.figure(2)
plt.semilogy(f,psdEta_1Sided)
plt.title('Power spectral density')
plt.xlabel('Frequency (Hz)')
plt.ylabel('Power/Frequency (m^2/Hz)')

#-----------------------------------------------------------------
```

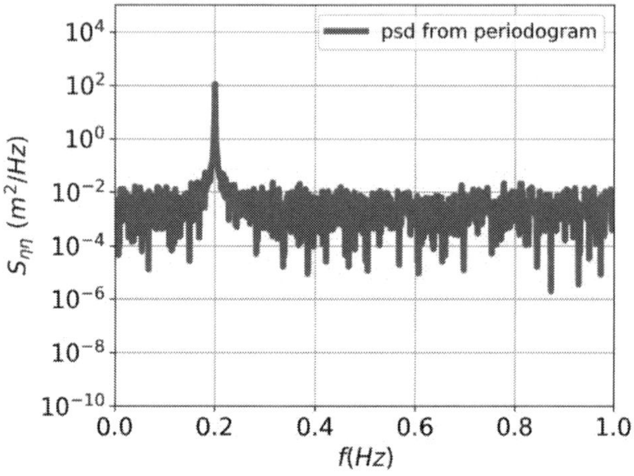

Figure 4.7 Demonstration of the power spectral density, *psd*, calculated by Python SciPy periodogram function. An analyzed time series has a primary frequency of $f = 0.2\ Hz$ and a sampling frequency of $f_s = 2\ Hz$. A resulted spectrum has a frequency range of $0 \leq f \leq f_s/2$ or $0 \leq f \leq 1$ with a peak wave frequency of $f_p = 0.2\ Hz$.

4.7 Calculating a power spectral density using MATLAB/Octave and Python/SciPy functions

Most of the programming languages have built-in functions to calculate a power spectral density. One widely used function to calculate a power spectral density is a periodogram. Using the same water surface time series generated above, a power spectral density is calculated by a periodogram function in MATLAB and GNU Octave as:

```
%Calculating a power spectral density (psd) using
MATLAB/Octave periodogram function
[psdEta_1Sided_PGram,fPGram]=periodogram(Eta,[],N,fs);
```

Now, results from a periodogram function are plotted as:

```
%Plotting a power spectral density from MATLAB/Octave
function
semilogy(fPGram,psdEta_1Sided_PGram,'b')
title('Power spectral density')
xlabel('Frequency (Hz)')
ylabel('Power/Frequency (m^2/Hz)')
legend('psd from periodogram')
```

Here is a code to calculate a power spectral density using Python SciPy periodogram function:

```
#------------------------------------------------------------------

#Calculating a power spectral density (psd) using Python (Scipy)
periodogram function
fPGram,psdEta_1Sided_PGram=sp.signal.periodogram(Eta,fs);

#------------------------------------------------------------------

#Plotting a power spectral density from Python function
plt.semilogy(fPGram,psdEta_1Sided_PGram,'b',label='psd from
periodogram')
plt.title('Power spectral density')
plt.xlabel('Frequency (Hz)')
plt.ylabel('Power/Frequency (m^2/Hz)')
pylab.legend(loc='upper right')

#------------------------------------------------------------------
```

4.8 Calculating wave properties from a spectral analysis method

In some research, values of the power spectrum and its shape are subjected to study. However, researchers mostly are interested in the properties of the signal obtained from a power spectral density. After a one-sided power spectral density is calculated from any of the methods mentioned above, wave properties are defined as:

$$m_n = \int_0^\infty S_{\eta\eta}(f) f^n df \approx \sum_{i=1}^N S_{\eta\eta,i} \times f_i^n \times \Delta f_i \tag{4.58}$$

$$m_0 = \int_0^\infty S_{\eta\eta}(f) df \approx \sum_{i=1}^N S_{\eta\eta,i} \times \Delta f_i = \sigma_\eta^2 \tag{4.59}$$

$$H_{m0} = 4\sigma_\eta = 4\sqrt{m_0} = 4\sqrt{\int_0^\infty S_{\eta\eta}(f)df} \approx 4\sqrt{\sum_{i=1}^{N} S_{\eta\eta,i} \times \Delta f_i} \tag{4.60}$$

$$T_p = \frac{1}{f_p} \tag{4.61}$$

$$T_{m01} = \frac{m_0}{m_1} \tag{4.62}$$

$$T_{m02} = \sqrt{\frac{m_0}{m_2}} \tag{4.63}$$

where, m_n is the nth moment of the power spectral density, m_0 is a zero-moment of the power spectral density, σ_η is a standard deviation of the water surface elevation, H_{m0} is zero-moment wave height, T_p is a peak wave period, f_p is a peak wave frequency which is a frequency associated with a maximum value of the $S_{\eta\eta}$, the T_{m01} is a mean wave period, T_{m02} is a mean zero-crossing period, and Δf is a frequency interval. A peak wave frequency is obtained by locating a frequency associated with a maximum value of the $S_{\eta\eta}$.

4.9 Wave properties calculation code

Now, a step by step guide is provided to calculate wave properties from a power spectral density using MATLAB or GNU Octave. Then, a Python code is presented.

To calculate wave properties, first, spectrum moments are calculated as:

```
%Calculating spectral moments
m0=sum(psdEta_1Sided.*f.^0.*df); %zero-moment of power
spectral density
m1=sum(psdEta_1Sided.*f.^1.*df); %first-moment of power
spectral density
m2=sum(psdEta_1Sided.*f.^2.*df); %second-moment of power
spectral density
```

Then, zero-moment wave height, H_{m0}, is calculated as:

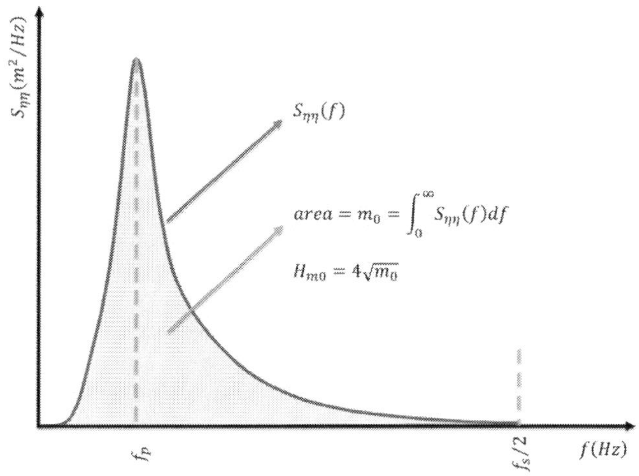

Figure 4.8 Schematic demonstration of the power spectral density with a frequency range of $0 \leq f \leq f_s/2$. A frequency associated with a maximum value of the spectrum is called a peak wave frequency, f_p. An area under a spectrum is equal to a zero-moment of the spectrum, m_0. The f_s is a sampling frequency, where $f_{Ny} = f_s/2$.

```
%Calculating wave height
Hm0=4*sqrt(m0); %zero-moment wave height
```

To calculate a peak wave period, first, a maximum value of the spectrum needs to be located. Commonly, a spectrum obtained from a periodogram method has rapid variations in its values. These rapid variations in a periodogram may cause an error in locating a true maximum value. Because of that, it is more suitable to smooth a power spectral density before locating a peak value. Smoothing a spectrum helps to prevent an error in locating a spectrum maximum value. A peak wave frequency, f_p, and a peak wave period, T_p, are calculated from a power spectral density (original one or smoothed one) as:

```
%Calculating a peak period
[psdEtaMax MaxIndx]=max(psdEta_1Sided(:,1)); %Locating an
index of the spectrum peak
fp=f(MaxIndx,1); %peak frequency
```

```
Tp=1/fp; %peak period
```

Finally, mean and zero crossing periods are calculated as:

```
%Calculating wave periods
Tm01=m0/m1; %mean period
Tm02=(m0/m2)^0.5; %zero crossing period
```

Here is a Python code to calculated wave properties from a power spectral density. Note that you need to install and import NumPy package as np along with Python.

```
#Calculating wave properties

#Calculating spectral moments
m0=np.sum(psdEta_1Sided*f**0*df); #zero-moment of power spectral
density
m1=np.sum(psdEta_1Sided*f**1*df); #first-moment of power spectral
density
m2=np.sum(psdEta_1Sided*f**2*df); #second-moment of power
spectral density

#Calculating wave height
Hm0=4*np.sqrt(m0); #zero-moment wave height

#Calculating a peak period
MaxIndx=np.argmax(psdEta_1Sided); #Locating an index of the
spectrum peak
fp=f[MaxIndx]; #peak frequency
Tp=1/fp; #peak period

#Calculating wave periods
Tm01=m0/m1; #mean period
Tm02=(m0/m2)**0.5; #zero crossing period
```

4.10 Peak wave frequency from weighted integral of wave power spectrum

Typically, a peak wave frequency, f_p, and a peak wave frequency, $T_p = 1/f_p$, are derived directly from a power spectral density. To do that, a frequency associated with a maximum value of $S_{\eta\eta}$ is located and considered as a peak wave frequency. Then, a peak wave period is calculated from a peak wave frequency. There is also another method to estimate a peak wave frequency from a spectrum. In this method, a peak wave frequency is estimated from a weighted integral of the power spectral density as (Young, 1995):

$$f_p = \frac{\int \left(S_{\eta\eta}(f)\right)^5 f df}{\int \left(S_{\eta\eta}(f)\right)^5 df} \tag{4.64}$$

where, f_p is a peak wave frequency estimated from a weighted integral of the power spectral density.

In general, it is suggested to obtain a peak wave frequency directly by locating it on a spectrum. Using a weighted integral is only recommended for conditions that there are rapid variations in $S_{\eta\eta}$, which does not allow to pick a maximum value of $S_{\eta\eta}$ accurately. Most of the time, these rapid variations can be eliminated by smoothing a spectrum. However, in some cases even smoothing might not help, such as a spectrum for low energy waves in shallow water.

Here is a MATLAB and GNU Octave code to calculate a peak wave frequency from a weighted integral:

```
%Calculating a peak wave frequency from a weighted integral
of the wave power spectrum

fp=(sum(psdEta_1Sided.^5.*f.^1*df))./(sum(psdEta_1Sided.^5.*
f.^0*df)); %peak frequency
```

Here is a Python code to calculate a peak wave frequency from a weighted integral. Note that you need to install and import NumPy package as np along with Python.

```
#Calculating a peak wave frequency from a weighted integral of the
wave power spectrum

fp=(np.sum(psdEta_1Sided**5*f**1*df))/(np.sum(psdEta_1Sided**5*f**
0*df)); #peak frequency
```

4.11 MATLAB, GNU Octave and Python

MATLAB (The MathWorks, Inc., www.mathworks.com) is a numerical computing environment, which is a very popular programming language with academic researchers. The GNU Octave (Eaton, 2016; www.gnu.org/software/octave) is an open source programming language that has a very compatible syntax with MATLAB. Python (Python Software Foundation; www.python.org) is a general purpose programming language, which is used in the development of the various applications. Python has a large number of scientific packages such as SciPy (Jones et al. 2001; www.scipy.org), NumPy (Van Der Walt et al., 2011; www.numpy.org), Pandas (McKinney, 2010; pandas.pydata.org), and Matplotlib (Hunter, 2007; matplotlib.org). These packages are developed by their community groups. The programming codes in this text are presented for MATLAB, GNU Octave, and Python. If these codes need to be used by any other programming language, then they need to be rewritten accordingly.

Anaconda distribution (Continuum Analytics, Inc; www.continuum.io) is used for Python package management. Spyder (github.com/spyder-ide/spyder) is used as an integrated development environment, IDE, for Python. Matplotlib (Hunter, 2007) is used as a primary plotting tool throughout this book. Here is a list of the software and packages that are used to develop codes in this book.

A good comparison between MATLAB and NumPy is presented at: docs.scipy.org/doc/numpy-dev/user/numpy-for-matlab-users.html. Some of the MATLAB and Python differences are explained below:

Tabel 4.1 List of the software and packages used in this book

Name	Description	Version
Matlab	Programming language	R2013a
Octave	Programming language	4.2.2
Python	Programming language	3.6.4
SciPy	Python package	1.0.0
NumPy	Python package	1.13.3
Matplotlib	Python package	2.1.2
Anaconda	Python distribution and package manager	5.0.1
Spyder	Python integrated development environment (IDE)	3.2.6

- MATLAB has a 1-based indexing or numbering. It means that an index of the first element in an array or a vector is 1. Python has a 0-based indexing, as an index of the first element in its array or its list is 0.
- MATLAB automatically grows a size of the arrays and vectors. However, if a larger array is required in Python, an array need to be resized.
- In MATLAB, the presence of the dot operator "." in front of other operators represents an element by the corresponding element operation. For example, $A.*B$ shows an element by element multiplication. In other words, $A.*B$ means that each element in an array A to be multiplied by a corresponding element in an array B. A dot operator "." is not required in Python.

Note that, in general, a single column array or vector is used in the codes of this text.

5 Window Functions, Digital Filters and Data Smoothing

In this chapter, window functions, digital filters, and data smoothing are covered. First, window functions are explained. Next, low-pass, high-pass, band-pass and band-reject filters are explained. Then, data smoothing procedure is described. Step by step sample codes to design a filter are presented.

5.1 Window function

A window function is a mathematical function that has a certain shape inside a given interval and typically is zero everywhere else outside that interval. Most of the window functions are a smooth curve with a bell shape. However, other shapes such as a rectangle and a triangle are also used. If a given function is multiplied by a window function within a finite interval, then a function is scaled proportional to a shape of the window function inside that interval and becomes zero outside of that interval (a window function is zero outside an interval). Some of the most common window functions are:

Rectangular window:

$$W(n_w) = 1 \tag{5.1}$$

Triangular or Bartlett window:

$$W(n_w) = 1 - \left| \frac{n_w - \frac{N_w - 1}{2}}{\frac{N_w - 1}{2}} \right| \tag{5.2}$$

Welch window:

$$W(n_w) = 1 - \left(\frac{n_w - \frac{N_w - 1}{2}}{\frac{N_w - 1}{2}} \right)^2 \tag{5.3}$$

Hann (Hanning) window:

$$W(n_w) = 0.5 - 0.5 \cos\left(\frac{2\pi n_w}{N_w - 1}\right) \tag{5.4}$$

Hamming window:

$$W(n_w) = 0.54 - 0.46 \cos\left(\frac{2\pi n_w}{N_w - 1}\right) \tag{5.5}$$

Gaussian window:

$$W(n_w) = \exp\left(-\frac{1}{2} \left(\frac{n_w - \frac{N_w - 1}{2}}{0.4 \times \frac{N_w - 1}{2}} \right)^2 \right) \tag{5.6}$$

Blackman window:

$$W(n_w) = 0.42659 - 0.49656 \cos\left(\frac{2\pi n_w}{N_w - 1}\right) + 0.076849 \cos\left(\frac{4\pi n_w}{N_w - 1}\right) \tag{5.7}$$

Nuttall window:

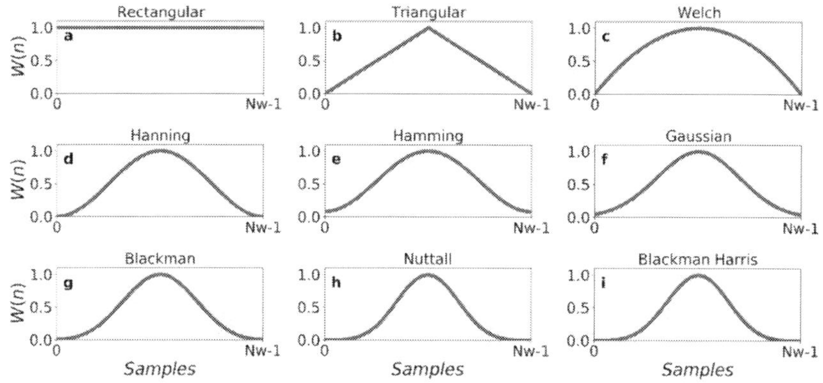

Figure 5.1 Window functions are demonstrated as (a): rectangular window, (b): triangular (Bartlett) window, (c): Welch window, (d): Hanning (Hann) window, I: Hamming window, (f): Gaussian window, (g): Blackman window, (h): Nuttall window, (i): Blackman Harris window. Window size is equal to N_w.

$$W(n_w) = 0.355768 - 0.487396\cos\left(\frac{2\pi n_w}{N_w - 1}\right) + 0.144232\cos\left(\frac{4\pi n_w}{N_w - 1}\right) - 0.012604\cos\left(\frac{6\pi n_w}{N_w - 1}\right) \tag{5.8}$$

Blackman–Harris window:

$$W(n_w) = 0.35875 - 0.48829\cos\left(\frac{2\pi n_w}{N_w - 1}\right) + 0.14128\cos\left(\frac{4\pi n_w}{N_w - 1}\right) - 0.01168\cos\left(\frac{6\pi n_w}{N_w - 1}\right) \tag{5.9}$$

where, $W(n_w)$ is a window function, N_w is a total number of points in a window function, i.e. a length of the window function, n_w is an element counter in a window function with a range of $0 \le n_w \le N_w - 1$.

5.2 Digital filters

A digital filter is a set of mathematical operations on the data in order to modify, decrease, or enhance certain aspects of the signal. Digital filters are used to modify signals in both time and frequency domains. However, sometimes a time domain dataset is modified indirectly by using a digital filter to manipulate a transformed dataset in the frequency domain. In that case, first, a time series is transformed to the frequency domain. Then, a

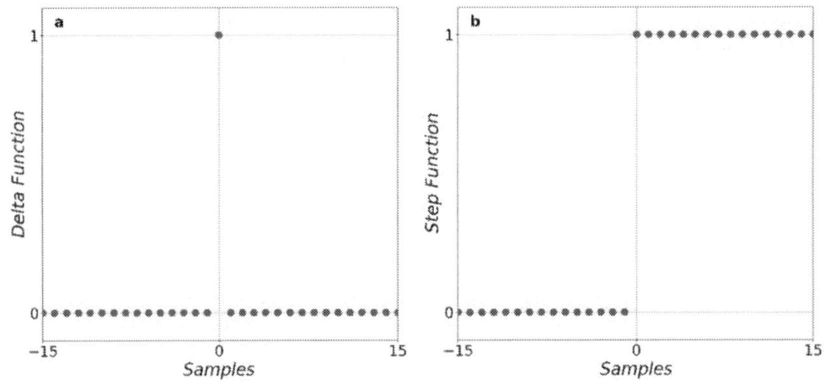

Figure 5.2 Plot (a) shows a delta function on a discrete domain. Plot (b) shows a unit step function on a discrete domain. A number of samples should be selected according to an intended analysis.

digital filter is used to modify signals in the frequency domain. After desired changes are applied, data are transformed back to the time domain. For example, rapid variations in a time series can be eliminated directly in the time domain. Also, it can be done in the frequency domain as well. To do that, first, a time series dataset is transformed into the frequency domain by using the Fourier transform. Next, high-frequency signals are removed from the Fourier transform. Then, modified values of the Fourier transform are transformed back to the time domain by using the inverse Fourier transform. By doing this, all rapid variations in the time domain are eliminated.

A digital filter is described by its impulse response and step response in the time domain and by its frequency response and phase response in the frequency domain. An impulse response, or an impulse response function, IRF, of the system is its output when a delta function, also known as an impulse, is used as an input signal. A delta function on a discrete domain is a normalized impulse where it is zero everywhere except at a single desired argument, commonly at zero, where it is equal to one. In other words, an impulse response describes a reaction of the system to a sudden change in input (e.g. Smith, 1997).

Step response of the system is its output when a step function is used as an input signal. Step response describes a reaction of the system to a step waveform, i.e. a unit step function, as an input. A unit step function on a discrete domain is a discontinuous function where it is zero for negative arguments, and is one for zero and positive arguments. Step response can obtain in two ways. First, feed a step function as an input to a system and an output is step response. Second, step response is equal to a running sum (discrete integration) of the impulse response.

A frequency response represents an impulse response in the frequency domain, and is obtained from the Fourier transform of the impulse response. A frequency response sometimes presents in decibels, dB, as:

$$dB = 20\log_{10} a = 10\log_{10} P \tag{5.10}$$

where, a is an amplitude and P is a power.

Digital filters are categorized as recursive and non-recursive. A recursive filter uses values both from an input signal and from its previous steps output as an input in a new step. An impulse response of the recursive filter is known as an infinite impulse response, IIR. In contrast, a non-recursive filter only uses values from an input signal, and do not use any values from its previous steps output. An impulse response of the non-recursive filter is known as a finite impulse response, FIR (e.g. Smith, 1997).

Among others, a low-pass filter, a high-pass filter, a band-pass filter, and a band-reject filter are the most common basic filters, which are discussed in following sections.

5.2.1 Low-pass filter

Assume that there is a desired high cut-off frequency, f_{cH} , where signal values with frequencies larger than f_{cH} need to be decreased or canceled, while signal values with frequencies smaller than f_{cH} remain untouched. To get a signal with this shape, a low-pass filter with a chosen f_{cH} is used. A low-pass filter is a digital filter that lets signal values with frequencies $f \leq f_{cH}$ pass while it attenuates signal values with $f > f_{cH}$. In this case, a passband of the filter is where $f \leq f_{cH}$ and a stopband of the filter is where $f > f_{cH}$. A filter bandwidth is a frequency range where a filter transition from a passband to a stopband takes place. In this filter, a cut-off frequency follows $f_{cH} \leq 0.5 f_s$.

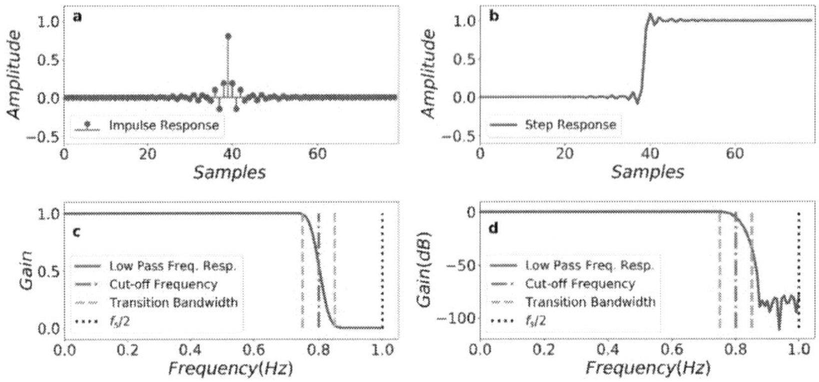

Figure 5.3 A low-pass windowed *sinc* filter. Plot (a) shows an impulse response of the low-pass filter with a high cut-off frequency of $f_{cH} = 0.8\,Hz$ and a bandwidth of $0.1\,Hz$. Plot (b) shows a step response of the low-pass filter presented in a plot (a). Plot (c) shows frequency response magnitudes or gains of the low-pass filter with a high cut-off frequency of $f_{cH} = 0.8\,Hz$, a bandwidth of $0.1\,Hz$, and a sampling frequency of $f_s = 2\,Hz$. Plot (d) shows frequency response magnitudes or gains presented in a plot (c) in Decibel, dB. An impulse response and a step response each has 79 data points. A frequency response has 129 data points. The Blackman window function is used in a filter calculation.

In the time domain, a simple low-pass filter is designed using a *sinc* function. To do that, first, a windowed *sinc* filter is calculated as:

$$h(n_w) = sinc\left(2\frac{f_{cL}}{f_s}\left(n_w - \frac{N_w - 1}{2}\right)\right) \times W(n_w) \tag{5.11}$$

where, $h(n_w)$ is a windowed *sinc* function filter, $W(n_w)$ is a window function, f_s is a sampling frequency in Hz, the N_w is a total number of points in a filter, i.e. a length of the filter, n_w is an element counter in a filter with a range of $0 \leq n_w \leq N_w - 1$, and $sinc(x) = \sin x/x$. A length of the filter, N_w, which should be an odd number, is defined based on a filter bandwidth as:

$$N_w = \frac{4}{(bw/f_s)} \tag{5.12}$$

where, bw is a filter bandwidth, and bw/f_s is a normalized filter bandwidth which must follow $bw/f_s \leq 0.5$. Then, a normalized low-pass windowed *sinc* filter, $h_{LowPassIR}$, in the time domain is calculated as:

$$h_{LowPassIR}(n_w) = \frac{h(n_w)}{\sum_{n_w=0}^{N_w-1} h(n_w)} \tag{5.13}$$

where, $h_{LowPassIR}$ is an impulse response of a low-pass windowed *sinc* filter. A summation of all elements in an impulse response of the low-pass windowed *sinc* filter is equal to one, i.e. $\sum_{n_w=0}^{N_w-1} h_{LowPassIR}(n_w) = 1$.

A frequency response of the low-pass filter is a magnitude (gain) of the low-pass filter in the frequency domain. A frequency response of the low-pass filter is obtained by transforming an impulse response of the filter to the frequency domain. This can be obtained from the Fourier transform of the low-pass filter impulse response. Before using the Fourier transform, if a length of the impulse response, N_w, is smaller than a length of the desired frequency response, N, (i.e. if $N_w < N$), then zero values are padded (added) at the end of the impulse response until a length of the impulse response becomes equal to N. Now, by using the Fourier transform, an impulse response of the filter is transformed to a frequency response of the filter in the frequency domain. These obtained values from the fast Fourier transform are complex numbers, consist of a magnitude (gain) and a phase. Mostly, magnitudes (gains) of the frequency response are used in the signal analysis.

5.2.2 High-pass filter

Assume that there is a low cut-off frequency,f_{cL} , where signal values with frequencies smaller than f_{cL} need to be decreased or canceled, while the signal values with frequencies larger than f_{cL} remain untouched. To get a signal with this shape, a high-pass filter with a chosen f_{cL} is used. A high-pass filter is a digital filter that lets signal values with frequencies $f \geq f_{cL}$ pass while it attenuates signal values with $f < f_{cL}$. In this case, a passband of the filter is where $f \geq f_{cL}$ and a stopband of the filter is where $f < f_{cL}$. A filter bandwidth is a frequency range where a filter transition from a passband to a stopband takes place. In this filter, a cut-off frequency follows $f_{cL} \leq 0.5 f_s$.

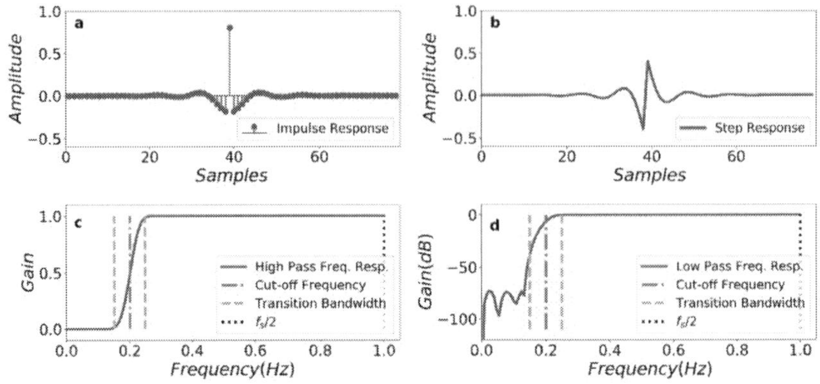

Figure 5.4 A high-pass windowed *sinc* filter. Plot (a) shows an impulse response of the high-pass filter with a low cut-off frequency of $f_{cL} = 0.2\ Hz$ and a bandwidth of $0.1\ Hz$. Plot (b) shows a step response of the high-pass filter presented in a plot (a). Plot (c) shows frequency response magnitudes or gains of a high-pass filter with a low cut-off frequency of $f_{cH} = 0.2\ Hz$, a bandwidth of $0.1\ Hz$, and a sampling frequency of $f_s = 2\ Hz$. Plot (d) shows frequency response magnitudes or gains presented in a plot (c) in Decibel, dB. An impulse response and a step response each has 79 data points. A frequency response has 129 data points. The Blackman window function is used in a filter calculation.

A simple high-pass filter is obtained by a spectral inversion of the low-pass filter. To do that, first, an impulse response of the low-pass windowed *sinc* filter, $h_{LowPassIR}$, is calculated by using a *sinc* function in the time domain. Then, a high-pass windowed *sinc* filter, $h_{HighPassIR}$, is calculated in the time domain as:

$$h_{HighPassIR}(n_w) = -h_{LowPassIR}(n_w) \tag{5.14}$$

$$h_{HighPassIR}(Center\ Point) = h_{HighPassIR}(Center\ Point) + 1 \tag{5.15}$$

where, $h_{HighPassIR}$ is an impulse response of a high-pass windowed *sinc* filter. A summation of all elements in an impulse response of the low-pass windowed *sinc* filter is equal to zero, i.e. $\sum_{n_w=0}^{N_w-1} h_{HighPassIR}(n_w) = 0$.

In other words, to obtain a high-pass windowed *sinc* filter, first, change a sign of all values in an impulse response of the low-pass windowed *sinc* filter, and then add one to a value of the element at a center of the filter.

A frequency response of the high-pass filter, which is a magnitude (gain) of the high-pass filter in the frequency domain, is obtained by transforming an impulse response of the high-pass filter to the frequency domain. This can be obtained from the Fourier transform of the high-pass filter impulse response. Before using the Fourier transform, if a length of the impulse response, N_w, is smaller than a length of the desired frequency response, N, (i.e. if $N_w < N$), then zero values are padded (added) at the end of the impulse response until a length of the impulse response becomes equal to N. Now, by using the Fourier transform, an impulse response of the filter is transformed to a frequency response of the filter in the frequency domain. These obtained values from the fast Fourier transform are complex numbers, consist of a magnitude (gain) and a phase. Mostly, magnitudes (gains) of the frequency response are used in the signal analysis.

Another method to obtain a high-pass filter magnitude (gain) in the frequency domain, i.e. a high-pass filter frequency response, is to change a sign of all values of the low-pass filter magnitude in the frequency domain, and then add one to all elements. In other words, a high-pass filter frequency response is equal to:

$$|h_{HighPassFR}(f)| = -|h_{LowPassFR}(f)| + 1 \qquad (5.16)$$

where, $h_{HighPassFR}$ is a frequency response of the high-pass windowed *sinc* filter, $|h_{HighPassFR}(f)|$ is a magnitude or gain of the high-pass frequency response, and f is a frequency, $h_{LowPassFR}$ is a frequency response of the low-pass windowed *sinc* filter, and $|h_{LowPassFR}(f)|$ is a magnitude or gain of the low-pass frequency response.

5.2.3 Band-pass filter

Assume that there is a low cut-off frequency, f_{cL} , and a high cut-off frequency, f_{cH} , where signal values with frequencies smaller than f_{cL} and larger than f_{cH} need to be decreased or canceled while the rest of the signal

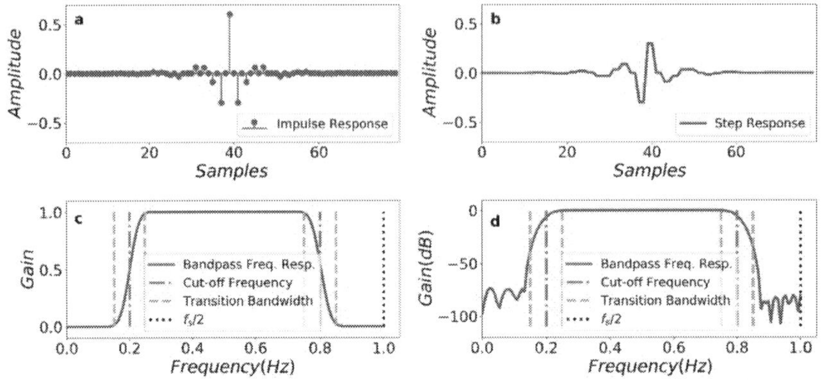

Figure 5.5 A band-pass windowed *sinc* filter. Plot (a) shows an impulse response of the band-pass filter with a low cut-off frequency of $f_{cL} = 0.2\ Hz$, a high cut-off frequency of $f_{cH} = 0.8\ Hz$, and a bandwidth of $0.1\ Hz$. Plot (b) shows a step response of the band-pass filter presented in a plot (a). Plot (c) shows frequency response magnitudes or gains of the band-pass filter with a low cut-off frequency of $f_{cL} = 0.2\ Hz$, a high cut-off frequency of $f_{cH} = 0.8\ Hz$, a bandwidth of $0.1\ Hz$, and a sampling frequency of $f_s = 2\ Hz$. Plot (d) shows frequency response magnitudes or gains presented in a plot (c) in Decibel, dB. An impulse response and a step response each has 79 data points. A frequency response has 129 data points. The Blackman window function is used in a filter calculation.

values remain untouched. In other words, signal values that have frequencies between f_{cL} and f_{cH} are passed. To get a signal with this shape, a band-pass filter is used. A band-pass filter is a digital filter that lets signal values with frequencies $f_{cL} \leq f \leq f_{cH}$ pass while it attenuates signal values with $f < f_{cL}$ or $f > f_{cH}$. In this case, a passband of the filter is where $f_{cL} \leq f \leq f_{cH}$ and a stopband of the filter is where $f < f_{cL}$ or $f > f_{cH}$. A filter bandwidth is a frequency range where a filter transition from a passband to a stopband takes place. In this filter, cut-off frequencies follow $f_{cL} < f_{cH}$, $f_{cH} \leq 0.5 f_s$ and $f_{cL} \leq 0.5 f_s$.

A simple band-pass filter is obtained by using low-pass and high-pass filters. To do that, first, an impulse response of the low-pass windowed *sinc* filter, $h_{LowPassIR}$, is calculated using f_{cH}. Next, an impulse response of the high-pass windowed *sinc* filter, $h_{HighPassIR}$, is calculated using f_{cL}. Then, a band-pass windowed *sinc* filter, $h_{BandPassIR}$, is calculated by convolving low-pass and high-pass filters in the time domain as:

$$h_{BandPassIR} = h_{LowPassIR\ (using\ f_{cH})} * h_{HighPassIR\ (using\ f_{cL})} \quad (5.17)$$

where, $h_{BandPassIR}$ is an impulse response of the band-pass windowed *sinc* filter. An asterisk sign, $*$, represents a convolution operation. A summation of all elements in an impulse response of the band-pass windowed *sinc* filter is equal to zero, i.e. $\sum_{n_w=0}^{N_w-1} h_{BandPassIR}(n_w) = 0$.

A frequency response of the band-pass filter, which is a magnitude (gain) of the band-pass filter in the frequency domain, is obtained by transforming an impulse response of the band-pass filter to the frequency domain. This can be obtained from the Fourier transform of the band-pass filter impulse response. Before using the Fourier transform, if a length of the impulse response, N_w, is smaller than a length of the desired frequency response, N, (i.e. if $N_w < N$), then zero values are padded (added) at the end of the impulse response until a length of the impulse response becomes equal to N. Now, by using the Fourier transform, an impulse response of the filter is transformed to a frequency response of the filter in the frequency domain. These obtained values from the fast Fourier transform are complex numbers, consist of a magnitude (gain) and a phase. Mostly, magnitudes (gains) of the frequency response are used in the signal analysis.

5.2.4 Band-reject filter

Assume that there is a low cut-off frequency, f_{cL} , and a high cut-off frequency, f_{cH} , where the signal values with frequencies between f_{cL} and f_{cH} need to be decreased or canceled while the rest of the signal values remain untouched. In other words, signal values that have frequencies smaller than f_{cL} and larger than f_{cH} are passed. To get a signal with this shape, a band-reject filter is used. A band-reject filter is a digital filter that lets signal values with frequencies $f \leq f_{cL}$ or $f \geq f_{cH}$ pass while it attenuates signal values if $f_{cL} < f < f_{cH}$. In this case, a passband of the filter is where $f \leq f_{cL}$ or $f \geq f_{cH}$ and a stopband of the filter is where $f_{cL} < f < f_{cH}$. A filter bandwidth is a frequency range where a filter transition from a passband to a stopband takes place. In this filter, cut-off frequencies follow $f_{cL} < f_{cH}$, $f_{cH} \leq 0.5 f_s$ and $f_{cL} \leq 0.5 f_s$.

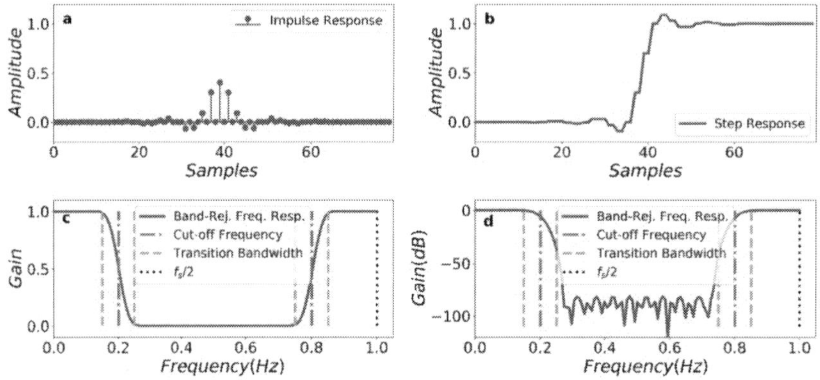

Figure 5.6 A band-reject windowed *sinc* filter. Plot (a) shows an impulse response of the band-reject filter with a low cut-off frequency of $f_{cL} = 0.2\ Hz$, a high cut-off frequency of $f_{cH} = 0.8\ Hz$, and a bandwidth of $0.1\ Hz$. Plot (b) shows a step response of the band-reject filter presented in a plot (a). Plot (c) shows frequency response magnitudes or gains of the band-reject filter with a low cut-off frequency of $f_{cL} = 0.2\ Hz$, a high cut-off frequency of $f_{cH} = 0.8\ Hz$, a bandwidth of $0.1\ Hz$, and a sampling frequency of $f_s = 2\ Hz$. Plot (d) shows frequency response magnitudes or gains presented in a plot (c) in Decibel, dB. An impulse response and a step response each has 79 data points. A frequency response has 129 data points. The Blackman window function is used in a filter calculation.

A simple band-reject filter is obtained by using low-pass and high-pass filters. To do that, first, an impulse response of the low-pass windowed *sinc* filter, $h_{LowPassIR}$, is calculated using f_{cL}. Next, an impulse response of the high-pass windowed *sinc* filter, $h_{HighPassIR}$, is calculated using f_{cH}. Then, a band-reject windowed *sinc* filter, $h_{BandRejectIR}$, is calculated by adding low-pass and high-pass filters together in the time domain as:

$$h_{BandRejectIR} = h_{LowPassIR\ (using\ f_{cL})} + h_{HighPassIR\ (using\ f_{cH})} \tag{5.18}$$

where, $h_{BandRejectIR}$ is an impulse response of a band-reject windowed *sinc* filter. A summation of all elements in an impulse response of the band-reject windowed *sinc* filter is equal to one, i.e. $\sum_{n_w=0}^{N_w-1} h_{BandRejectIR}(n_w) = 1$.

A frequency response of the band-reject filter, which is a magnitude (gain) of the band-reject filter in the frequency domain, is obtained by transforming an impulse response of the band-reject filter to the frequency domain. This can be obtained from the Fourier transform of the band-reject filter impulse response. Before using the Fourier transform, if a length of the impulse response, N_w, is smaller than a length of the desired frequency response, N, (i.e. if $N_w < N$), then zero values are padded (added) at the end of the impulse response until a length of the impulse response becomes equal to N. Now, by using the Fourier transform, an impulse response of the filter is transformed to a frequency response of the filter in the frequency domain. These obtained values from the fast Fourier transform are complex numbers, consist of a magnitude (gain) and a phase. Mostly, magnitudes (gains) of the frequency response are used in the signal analysis.

5.2.5 Low-pass filter code

Now, a step by step practical guide is provided on how to generate a low-pass windowed *sinc* filter using MATLAB or GNU Octave. Then, a Python code is presented. First, filter specifications are given as:

```
%Filter specifications
fs=2; %Sampling frequency (Hz)
fcH=0.8; %High cut-off frequency (Hz)
BandWidth=0.1; %Filter bandwidth
N=256; %Length of the frequency response
```

Now, a frequency vector containing $\left(\frac{N}{2}+1\right)$ frequency values in a range of $0 \leq f \leq f_s/2$, where f_s is a sampling frequency, f_{Ny} is the Nyquist frequency ($f_s/2 = f_{Ny}$), and N is a total number of samples in a frequency response, is generated as:

```
%Generating a frequency vector
df=fs/N; %Frequency difference between consecutive samples,
df=fs/N
f(:,1)=[0:df:fs/2]; %Frequency vector with (N/2+1) elements
from 0 Hz to fNy=fs/2 Hz, equally spaced at df
```

A length of the low-pass filter, N_w, is defined by implementing a normalized bandwidth, i.e. bw/f_s, in an equation as $N_w = \frac{4}{(bw/f_s)}$. A length of the low-pass filter, N_w, must be an odd number and equal or larger than 3.

```
%Defining a length of the filter
BandWidthNorm=BandWidth/fs; %Normalized transition bandwidth
respect to a sampling frequency: between 0 to 0.5 Hz
Nw=ceil(4/BandWidthNorm); %Transition bandwidth size, equal
to a number of the elements in window and sinc functions,
must be odd
Nw(mod(Nw,2)==0)=Nw-1; %BandWidthSize must be odd
```

Create a proper window function. Here, the Blackman window function is used.

```
%Generating the Blackman window function
nwin(:,1)=[0:1:Nw-1]; %Counter from 0 to Nw-1
Wn=0.42659-0.49656*cos(2*pi*nwin/(Nw-
1))+0.076849*cos(4*pi*nwin/(Nw-1)); %Blackman window
function
```

Create a windowed *sinc* function as:

```
%sinc function from 0 to Nw-1
SincFun=sinc(2*(fcH/fs)*(nwin-(Nw-1)/2));

%windowed-sinc function
WnSinc=Wn.*SincFun; %windowed-sinc function
```

An impulse response of the windowed *sinc* function is calculated by normalizing a windowed *sinc* function. To do that, windowed *sinc* function elements are divided by a total summation of the elements, i.e. $h_{LowPassIR}(n_w) = \frac{h(n_w)}{\sum_{n_w=0}^{N_w-1} h(n_w)}$.

```
LowPassImpulseResponse=WnSinc./sum(WnSinc(:,1)); %Impulse
response of the windowed-sinc function in the time domain
```

Now, a frequency response of the filter is calculated. This is commonly obtained from the Fourier transform of the impulse response. First, a length of the impulse response in the time domain is compared with a length of the desired frequency response in the frequency domain. If a length of the impulse response, N_w, is smaller than a length of the desired frequency response, N, (i.e. if $N_w < N$), then zero values are padded (added) at the end of the impulse response until a length of the impulse response becomes equal to N.

```
%Padding zero to the end of the impulse response to make its
length equal to N
LowPassImpulseResponseWithZeroPadding(1:length(nwin(:,1)),1)
=LowPassImpulseResponse;
LowPassImpulseResponseWithZeroPadding(length(nwin(:,1))+1:N,
1)=0;
```

Now, by using the fast Fourier transform, an impulse response of the low-pass filter is transformed to a frequency response of the filter in the frequency domain. These obtained values from the fast Fourier transform are complex numbers, consist of a magnitude (gain) and a phase. Mostly, magnitudes (gains) of the frequency response are used in the signal analysis.

```
LowPassFrequencyResponse=fft(LowPassImpulseResponseWithZeroP
adding); %windowed-sinc filter
LowPassFrequencyResponseGain=abs(LowPassFrequencyResponse);
%windowed-sinc filter magnitude (gain)
LowPassFrequencyResponsePhase=angle(LowPassFrequencyResponse
); %windowed-sinc filter phase
```

A frequency response calculated in the last step has a frequency range of $0 \leq f \leq f_s - \Delta f$, where Δf is a frequency interval. For a practical purpose, only half of the filter is used with a range of $0 \leq f \leq f_s/2$ as:

```
LowPassFrequencyResponseGain=LowPassFrequencyResponseGain(1:
length(f(:,1))); %(N/2+1)elements if N is even, and (N+1)/2
elements if N is odd
```

```
LowPassFrequencyResponsePhase=LowPassFrequencyResponsePhase(
1:length(f(:,1))); %(N/2+1)elements if N is even, and
(N+1)/2 elements if N is odd
```

Magnitudes, i.e. gains, of the frequency response in Decibel are:

```
LowPassFrequencyResponseGain_dB=20.*log10(abs(LowPassFrequen
cyResponseGain)); %Converting to Decibel
```

A step response of the low-pass filter is calculated from a running sum (a discrete integration) of the impulse response as:

```
% Calculating a step response for a low-pass filter
LowPassStepResponse=zeros(length(LowPassImpulseResponse(:,1)
),1);
LowPassStepResponse(1,1)=LowPassImpulseResponse(1,1);
for i=2:length(LowPassImpulseResponse(:,1))
    LowPassStepResponse(i,1)=LowPassStepResponse(i-
1,1)+LowPassImpulseResponse(i,1);
end
```

Filter responses in the time and frequency domains are plotted as:

```
%Plotting

subplot(2,2,1)
stem(nwin,LowPassImpulseResponse)
title('Impulse Response')
xlabel('Samples')
ylabel('Amplitude')

subplot(2,2,2)
plot(nwin,LowPassStepResponse)
title('Step Response')
xlabel('Samples')
ylabel('Amplitude')

subplot(2,2,3)
plot(f,LowPassFrequencyResponseGain);
title('Low Pass Frequency Response')
xlabel('Frequency (Hz)')
ylabel('Gain')
```

```
subplot(2,2,4)
plot(f,LowPassFrequencyResponseGain_dB);
title('Low Pass Frequency Response')
xlabel('Frequency (Hz)')
ylabel('Gain (dB)')
```

Here is a Python code for calculating a low-pass windowed *sinc* filter. Note that you need to install and import NumPy package as np and matplotlib.pyplot package as plt along with the Python.

```
#-----------------------------------------------------------------
#Importing required packages

import numpy as np
import matplotlib.pyplot as plt

#-----------------------------------------------------------------
#Filter specifications
fs=2; #Sampling frequency (Hz)
fcH=0.8; #High cut-off frequency (Hz)
BandWidth=0.1; #Filter bandwidth
N=256; #Length of the frequency response

#-----------------------------------------------------------------
#Generating a frequency vector
df=fs/N; #Frequency difference between consecutive samples,
df=fs/N
f=np.arange(0,fs/2+df,df); #Frequency vector with (N/2+1) elements
from 0 Hz to fNy=fs/2 Hz, equally spaced at df

#-----------------------------------------------------------------
#Creating a low-pass filter

#Defining a length of the filter
BandWidthNorm=BandWidth/fs; #Normalized transition bandwidth
respect to a sampling frequency: between 0 to 0.5 Hz
Nw=int(np.ceil(4/BandWidthNorm)); #Transition bandwidth size,
equal to a number of the elements in window and sinc functions,
must be odd
if ((Nw%2)==0): Nw=Nw-1; #Bandwidth size must be odd

#Generating the Blackman window function
nwin=np.arange(0,Nw,1); #Counter from 0 to Nw-1
Wn=0.42659-0.49656*np.cos(2*np.pi*nwin/(Nw-
1))+0.076849*np.cos(4*np.pi*nwin/(Nw-1)); #Blackman window
function

#sinc function from 0 to Nw-1
SincFun=np.sinc(2*(fcH/fs)*(nwin-(Nw-1)/2));
```

```
#windowed-sinc function
WnSinc=Wn*SincFun; #windowed-sinc function
LowPassImpulseResponse=WnSinc/np.sum(WnSinc); #Impulse response of
the windowed-sinc function in the time domain

#-----------------------------------------------------------------
#Calculating a frequency response for a low-pass filter

#Padding zero to the end of the impulse response to make its
length equal to N
LowPassImpulseResponseWithZeroPadding=np.zeros(N);
LowPassImpulseResponseWithZeroPadding[0:len(nwin)]=LowPassImpulseR
esponse;

LowPassFrequencyResponse=np.fft.fft(LowPassImpulseResponseWithZero
Padding); #windowed-sinc filter
LowPassFrequencyResponseGain=np.abs(LowPassFrequencyResponse);
#windowed-sinc filter magnitude (gain)
LowPassFrequencyResponsePhase=np.angle(LowPassFrequencyResponse);
#windowed-sinc filter phase

LowPassFrequencyResponseGain=LowPassFrequencyResponseGain[0:len(f)
]; #(N/2+1)elements if N is even, and (N+1)/2 elements if N is odd
LowPassFrequencyResponsePhase=LowPassFrequencyResponsePhase[0:len(
f)]; #(N/2+1)elements if N is even, and (N+1)/2 elements if N is
odd

LowPassFrequencyResponseGain_dB=20*np.log10(np.abs(LowPassFrequenc
yResponseGain)); #Converting to Decibel

#-----------------------------------------------------------------
#Calculating a step response for a low-pass filter

LowPassStepResponse=np.zeros(len(LowPassImpulseResponse))
LowPassStepResponse[0]=LowPassImpulseResponse[0]
for i in range(1,len(LowPassImpulseResponse),1):
    LowPassStepResponse[i]=LowPassStepResponse[i-
1]+LowPassImpulseResponse[i]

#-----------------------------------------------------------------
#Plotting

plt.subplot(2,2,1)
plt.stem(nwin,LowPassImpulseResponse,label='Impulse Response')
plt.title('Impulse Response')
plt.xlabel('Samples')
plt.ylabel('Amplitude')

plt.subplot(2,2,2)
plt.plot(nwin,LowPassStepResponse,label='Step Response')
plt.title('Step Response')
plt.xlabel('Samples')
```

```
plt.ylabel('Amplitude')

plt.subplot(2,2,3)
plt.plot(f,LowPassFrequencyResponseGain,label='Low Pass Freq.
Resp.');
plt.title('Low Pass Frequency Response')
plt.xlabel('Frequency (Hz)')
plt.ylabel('Gain')

plt.subplot(2,2,4)
plt.plot(f,LowPassFrequencyResponseGain_dB,label='Low Pass Freq.
Resp.');
plt.title('Low Pass Frequency Response')
plt.xlabel('Frequency (Hz)')
plt.ylabel('Gain (dB)')

#-----------------------------------------------------------------
```

5.2.6 High-pass filter code

A high-pass windowed *sinc* filter is obtained from a low-pass windowed *sinc* filter. Assuming a low-pass filter is already available, a step by step practical guide is provided on how to calculate a high-pass windowed *sinc* filter using MATLAB or GNU Octave. Then, a Python code is presented. A high-pass filter from a low-pass filter is calculated as:

```
%Creating a high-pass filter from a low-pass filter
HighPassImpulseResponse=-LowPassImpulseResponse; %Impulse
response of the high-pass filter in the time domain
HighPassImpulseResponse((Nw-
1)/2+1,1)=HighPassImpulseResponse((Nw-1)/2+1,1)+1; %Impulse
response of the high-pass filter in the time domain

HighPassFrequencyResponseGain=-
LowPassFrequencyResponseGain+1; %High-pass filter magnitude
(gain)

HighPassFrequencyResponseGain_dB=20.*log10(abs(HighPassFrequ
encyResponseGain)); %Converting to Decibel
```

A step response of the high-pass filter is calculated from a running sum (a discrete integration) of the impulse response as:

```
%Calculating a step response for a high-pass filter
HighPassStepResponse=zeros(length(HighPassImpulseResponse(:,
1)),1);
HighPassStepResponse(1,1)=HighPassImpulseResponse(1,1);
for i=2:length(HighPassImpulseResponse(:,1))
    HighPassStepResponse(i,1)=HighPassStepResponse(i-
1,1)+HighPassImpulseResponse(i,1);
End
```

Here is a Python code for calculating a high-pass windowed *sinc* filter. Note that you need to install and import NumPy package as np and matplotlib.pyplot package as plt along with the Python.

```
#----------------------------------------------------------------
#Creating a high-pass filter from a low-pass filter
HighPassImpulseResponse=-LowPassImpulseResponse; #Impulse response
of the high-pass filter in the time domain
HighPassImpulseResponse[int((Nw-
1)/2)]=HighPassImpulseResponse[int((Nw-1)/2)]+1; #Impulse response
of the high-pass filter in the time domain

HighPassFrequencyResponseGain=-LowPassFrequencyResponseGain+1;
#High-pass filter magnitude (gain)

HighPassFrequencyResponseGain_dB=20*np.log10(np.abs(HighPassFreque
ncyResponseGain)); #Converting to Decibel

#----------------------------------------------------------------
#Calculating a step response for a high-pass filter
HighPassStepResponse=np.zeros(len(HighPassImpulseResponse))
HighPassStepResponse[0]=HighPassImpulseResponse[0]
for i in range(1,len(HighPassImpulseResponse),1):
    HighPassStepResponse[i]=HighPassStepResponse[i-
1]+HighPassImpulseResponse[i]

#----------------------------------------------------------------
```

After low-pass and high-pass filters are calculated, a pass-band or a pass-reject filter can be calculated from them. The website https://fiiir.com (from TomRoelandts.com) provides an interactive simple filter design.

5.3 Filter application in the time domain

Any of these discussed filters can be applied either in the time domain or in the frequency domain. To apply these filters in the time domain, a convolution operation is used. In mathematics, a convolution is defined as an integral of the product of two functions F and G after one is reversed or mirrored and shifted across the other one. The result is the third function, which is a modified version for one of the original functions. A convolution of F and G is:

$$(F * G)(t) = \int_{-\infty}^{\infty} F(\tau)G(t-\tau)d\tau \tag{5.19}$$

In a discrete form, a convolution of F and G is:

$$(F * G)(n) = \sum_{m=-M}^{M} F(n-m)G(m) \tag{5.20}$$

where, t and τ are representing variables in functions F and G, and an asterisk sign, $*$, represents a convolution operation. Note that t and τ can be any variable as long as they both represent the same variable. For example, in the time domain they both represent time, in the space domain they both represent a distance, and in the frequency domain they both represent a frequency. Consider F as an original function and G as a weighting function. Then, a convolution $(F * G)(t)$ gives a weighted average of the function $F(\tau)$ at the t, where the weightings given by function $G(\tau)$ are simply shifted by an amount of t. As t varies, a weighting function G emphasizes on different parts of the input function F.

Now, to filter data in the time domain (to apply a digital filter on data in the time domain), a convolution is used by considering F as an original dataset and G as an impulse response of the filter. Then, a dataset F is filtered in the time domain as:

$$F_{Filtered} = \left(F * h_{Imluse\ Response}\right)(n) \tag{5.21}$$

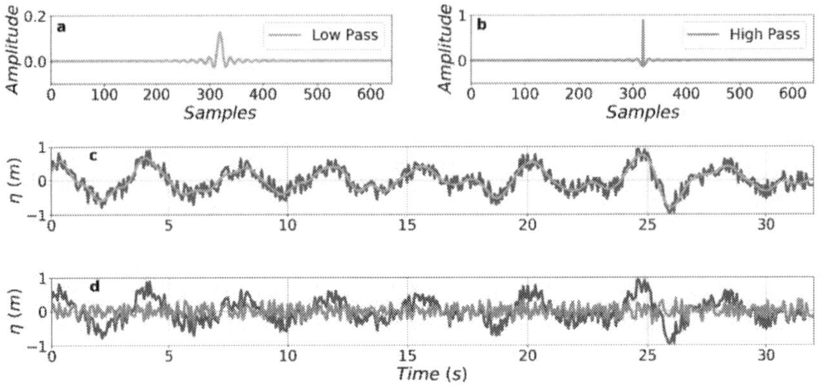

Figure 5.7 Filtering the time series data, η. Plot (a) shows an impulse response of a low-pass filter with a high cut-off frequency of $f_{cH} = 1\ Hz$ and a bandwidth of $0.1\ Hz$. Plot (b) shows an impulse response of a high-pass filter with a low cut-off frequency of $f_{cL} = 1\ Hz$ and a bandwidth of $0.1\ Hz$. Plot (c) shows a time series of the original and the low-pass filtered values. Plot (d) shows a time series of the original and the high-pass filtered values. A filtered time series is calculated by convolving an original time series with an impulse response. A time series represent the JONSWAP spectrum (Hasselmann et al., 1973) for a wind velocity of $10\ m/s$, a wind fetch of $10000\ m$, and a peak wave frequency of $0.25\ Hz$, contaminated with randomly generated noise between $-0.3\ m$ and $0.3\ m$. A time series has a duration of $1024\ s$ and a sampling frequency of $f_s = 16\ Hz$. Only 32 seconds of the time series is plotted.

5.4 Filter application in the frequency domain

In mathematics, a convolution in the time domain is equivalent to a pointwise multiplication in the frequency domain. It means that to use a digital filter in the frequency domain, an input signal is transformed to the frequency domain by using the Fourier transform, and then it is simply multiplied by a frequency response of the filter, $h_{Frequency\ Response}$. Therefore, a dataset F is filtered in the frequency domain (a digital filter is applied on a dataset F in the frequency domain) as:

$$F_{Filtered} = \mathcal{F}(F) \times h_{Frequency\ Response} \tag{5.22}$$

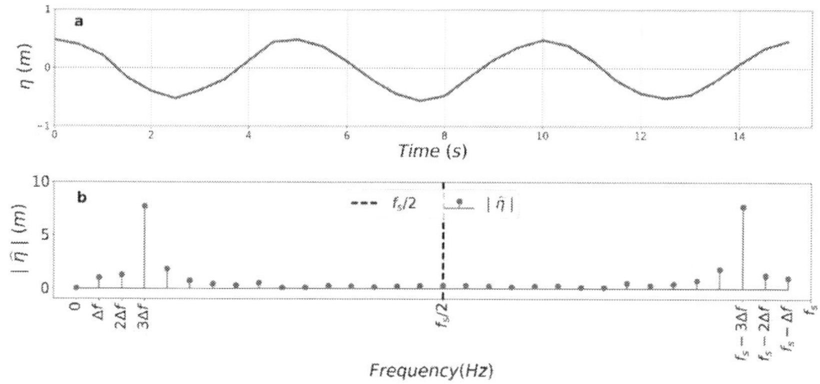

Figure 5.8 Symmetrical magnitude of the Fourier transform data (except for $f = 0$). The $\hat{\eta}$ is the Fourier transform and $|\hat{\eta}|$ is a magnitude of the Fourier transform. Plot (a) shows a time series of the wave. Plot (b) shows magnitudes of the two-sided Fourier transform with a frequency range of $0 \leq f \leq f_s - \Delta f$. Except for $f = 0\ Hz$, rest of the Fourier transform is symmetrical with respect to $f_s/2$. A time series and the Fourier transform represent a cosine wave with an amplitude of $0.5\ m$ and a period of $5\ s$, contaminated with randomly generated noise between $-0.1\ m$ and $0.1\ m$. A time series has a duration of $16\ s$ and a sampling frequency of $f_s = 2\ Hz$. Only 15.5 seconds of the time series is plotted.

where, $F_{Filtered}$ shows filtered values of F in the frequency domain, F is a dataset in the time domain, $\mathcal{F}(F)$ is the Fourier transform of the F, and $h_{Frequency\ Response}$ is a frequency response of the filter.

About this method, it needs to be pointed out that most of the signal analysis packages return a one-sided frequency response with a frequency range of $0 \leq f \leq f_s/2$. On the other hand, programming languages commonly return two-sided Fourier transform values with a frequency range of $0 \leq f \leq f_s - \Delta f$, where Δf is a frequency interval. In that case, either a one-sided frequency response needs to be extended to become two-sided, or a two-sided Fourier transform needs to be shortened to become one-sided. If all the analysis is intended to be done in the frequency domain, then one-sided values are preferable. Otherwise, if filtered data are intended to be transformed back to the time domain, then two-sided values are suitable.

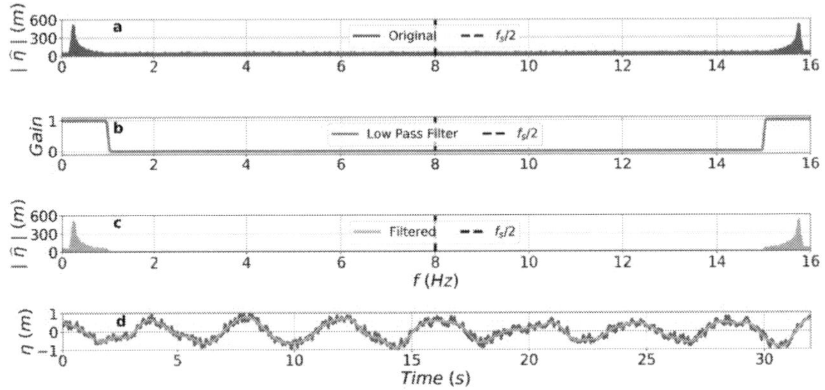

Figure 5.9 Filtering data in the frequency domain. The $\hat{\eta}$ is the Fourier transform and $|\hat{\eta}|$ is a magnitude of the Fourier transform. Plot (a) shows magnitudes of the Fourier transform of the original time series. Plot (b) shows a two-sided frequency response of a low-pass filter with a high cut-off frequency of $f_{cH} = 1\ Hz$ and a bandwidth of 0.1 Hz. Plot (c) shows magnitudes of the filtered Fourier transform data. To obtain that, the Fourier transform data are multiplied pointwise by frequency response magnitudes (gain values). Plot (d) shows a time series of the original and filtered values. A filtered time series is calculated from an inverse Fourier transform of the filtered Fourier transform. Note that, values in plots (a, b and c) are mirrored with respect to $f_s/2$ (except for $f = 0$). A time series and the Fourier transform represent the JONSWAP spectrum (Hasselmann et al., 1973) for a wind velocity of 10 m/s, a wind fetch of 10000 m, and a peak wave frequency of 0.25 Hz, contaminated with randomly generated noise between $-0.3\ m$ and $0.3\ m$. A time series has a duration of 1024 s and a sampling frequency of $f_s = 16\ Hz$. Only 32 seconds of the time series is plotted.

Programming languages mostly use the fast Fourier transform algorithm to calculate the discrete Fourier transform. For a time series consists of real values, the fast Fourier transform outcome is symmetrical with respect to $f_s/2$, except for $f = 0$. Therefore, a conversion of the one-sided frequency response to a two-sided frequency response can be done by mirroring a frequency response data. To do that, first, a frequency response values with a frequency range of $0 < f < f_s/2$ are mirrored with respect to $f_s/2$. Do not include $f = 0$ and $f = f_s/2$ in the mirroring procedure. Then, these mirrored values which have a frequency range of $f_s/2 + \Delta f \leq f \leq f_s - \Delta f$ are added to existing frequency response values which have a frequency range of $0 \leq f \leq f_s/2$. The outcome is a two-sided frequency response, which is symmetrical (except for $f = 0$) with respect to $f_s/2$, and has a frequency range of $0 \leq f \leq f_s - \Delta f$.

5.5 Filtering a power spectral density

In spectral analysis, it is usual to only retain a section of the spectrum and discard the rest. In general, this process is achieved by using a digital filter such as a low-pass, a high-pass, or a band-pass filter. However, in the coastal and ocean wave data analysis, it is common to modify a one-sided spectrum directly, without using a digital filter. For example, instead of using a mathematical filter to remove an undesired section of the spectrum, in practice, an undesired section is simply deleted from a spectrum.

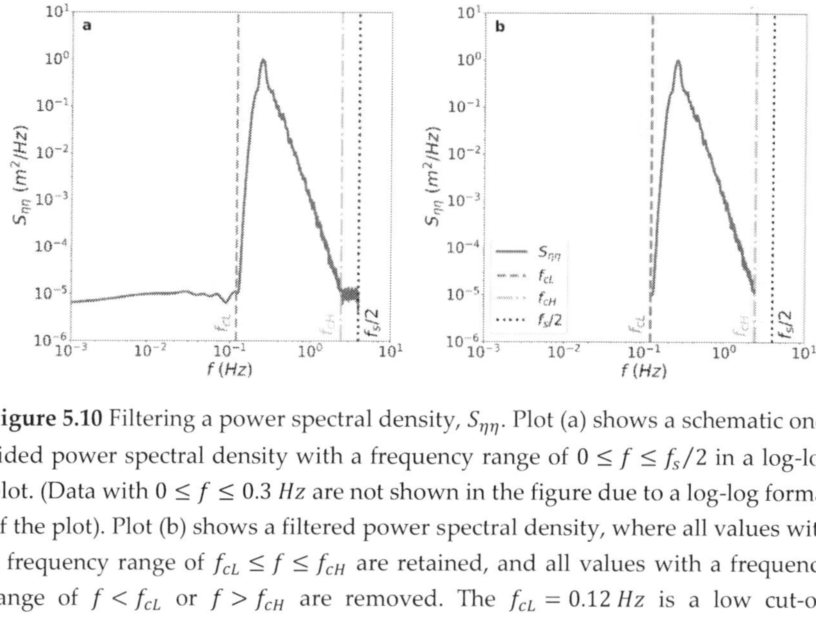

Figure 5.10 Filtering a power spectral density, $S_{\eta\eta}$. Plot (a) shows a schematic one-sided power spectral density with a frequency range of $0 \leq f \leq f_s/2$ in a log-log plot. (Data with $0 \leq f \leq 0.3$ Hz are not shown in the figure due to a log-log format of the plot). Plot (b) shows a filtered power spectral density, where all values with a frequency range of $f_{cL} \leq f \leq f_{cH}$ are retained, and all values with a frequency range of $f < f_{cL}$ or $f > f_{cH}$ are removed. The $f_{cL} = 0.12$ Hz is a low cut-off frequency, $f_{cH} = 2.4$ Hz is a high cut-off frequency, and $f_s = 8$ Hz is a sampling frequency. Values of f_{cL}, f_{cH} and f_s in this figure are selected for a demonstration purpose. In practice, these values should be selected based on physical properties of the study area, instrumentation, and nature of the research. A schematic power spectral density is obtained from the manipulated JONSWAP spectrum (Hasselmann et al., 1973) for a wind velocity of 10 m/s, a wind fetch of 10000 m, and a peak wave frequency of 0.25 Hz, contaminated with randomly generated noise.

Here are some examples of how to modify a spectrum directly. As the first example, assume there is a dataset that has low-frequency swell energies in its spectrum. Now, if these low-frequency swell energies need to be removed from a spectrum, first, a low cut-off frequency, f_{cL} , is defined. Then, all values in a spectrum that have frequencies smaller than a cut-off frequency, i.e. $f < f_{cL}$, are removed from a spectrum. As another example, assume there is a dataset that shows unacceptable noise in a high-frequency tail of its spectrum. These unacceptable high-frequency values need to be removed from a spectrum. To do that, a high cut-off frequency, f_{cH}, is defined, and all values in a spectrum that have frequencies larger than that a cut-off frequency, i.e. $f > f_{cH}$, are removed from a spectrum. A combination of two previous examples can occur if a certain frequency range of the spectrum located between a low cut-off frequency and a high cut-off frequency, i.e. $f_{cL} \leq f \leq f_{cH}$, are retained, while the rest of the data in a spectrum are deleted.

5.6 Data smoothing

In the data analysis, it is usual that original data or results demonstrate rapid variations and fluctuations. Also, data or results might be contaminated with noise. This is particularly true for data that are collected in the real world. For example, a periodogram usually shows lots of rapid variations and makes it difficult to follow the general trend of the spectrum. In such cases that a dataset shows lots of rapid fluctuations, a smoothing technique is used to capture important patterns in a dataset while leaving out rapid variations such as noise. In a smoothing procedure, all data points that are away from the main trend of the data are replaced by the ones that resemble the general trend in data. There are sophisticated algorithms and methods for smoothing the data. However, one simple way to smooth the data is to use a convolution. This can be done in any domain, such as the time or frequency domains. Make sure that a convolution application for smoothing is not confused with the one used for a digital filter (read a note on their differences at the end of this section). Considering that a convolution gives a weighted average of the original function, the original data can be smoothed by a properly chosen window function. For example, a power spectral density of the water surface elevation in the frequency domain can be smoothed by the Hamming window function as:

$$W(n_w) = 0.54 - 0.46\cos\left(\frac{2\pi n_w}{N_w - 1}\right) \tag{5.23}$$

$$W_N(n_w) = \frac{W(n_w)}{\sum W(n_w)} \tag{5.24}$$

$$S_{\eta\eta-Smoothed}(f) = S_{\eta\eta}(f) * W_N \tag{5.25}$$

where, $W(n_w)$ is a window function (here, the Hamming window function), $W_N(n_w)$ is a normalized window function in such a way that $\sum W_N(n_w) = 1$, the $S_{\eta\eta}$ is a one-sided water surface elevation power spectral density, $S_{\eta\eta-Smoothed}$ is a smoothed one-sided water surface elevation power spectral density, f is a frequency, and an asterisk sign, $*$, represents a convolution operation. A smoothing procedure should not change an area under a spectrum. In other words, $\int_0^\infty S_{\eta\eta}(f)df = \int_0^\infty S_{\eta\eta-Smoothed}(f)df$.

If a rectangular window function, i.e. $W(n_w) = 1$, is chosen for a convolution operation, then obtained results from a convolution is equal to a moving average of the data. For example, a water surface elevation in the time domain can be smoothed by a rectangular window function as:

$$W(n_w) = 1 \tag{5.26}$$

$$W_N(n_w) = \frac{W(n_w)}{\sum W(n_w)} \tag{5.27}$$

$$\eta_{Smoothed}(t) = \eta(t) * W_N \tag{5.28}$$

where, η is a water surface fluctuation in the time domain, and t is time.

It worth pointing out that using a convolution for smoothing the data should not be mistaken by using a convolution to apply a digital filter. To apply a digital filter in the time domain, a time series is convolved with an impulse response of the filter. To apply a digital filter in the frequency domain, the Fourier transform of the time series is multiplied by a frequency response of the filter (a pointwise multiplication). Therefore, for a digital filter, an effect obtained from a multiplication operation in the frequency domain is exactly the same as the one generated by a convolution operation in the time domain.

However, for a smoothing process, a convolution operation itself is directly used in either of the time or frequency domains. It means that, to smooth the data in the time domain, a time series is convolved with a proper window function. Similarly, to smooth a signal in the frequency domain, regardless of its relationship with the time domain, a signal is convolved with a proper window function (in the frequency domain).

5.7 Smoothing a power spectral density (periodogram)

A power spectral density or a periodogram typically contains values with rapid variations and fluctuations. Because of that, it is common to smooth a spectrum to show a trend of the spectrum values more clearly. There are sophisticated methods that can be used to smooth a power spectrum. Some methods are applied on data before using the Fourier transform, while some applied on the spectrum itself.

One method to obtain a smoothed power spectral density or an averaged periodogram is the Bartlett's method (Bartlett, 1948, 1950). In this method, first, time series data are divided into a series of non-overlapping segments. Then, a periodogram for each of these segments is calculated separately. Finally, a smoothed periodogram is obtained by averaging all periodograms calculated for these segments (e.g. Djuric et al., 1999; Gupta et al., 2013; Rahi and Mehra; 2014; Karuna and Mahalakshmi, 2015).

One of the most common methods for smoothing a periodogram is the Welch's method (Welch, 1967). In this method, first, time series data are divided into a series of overlapping segments. Then, each of these overlapping segments is windowed using a window function. Then, a periodogram for each of these segments is calculated separately. Most window functions leave a center of each segment untouched while suppressing the data points at both end of that segment. This causes an energy loss which needs to be compensated by dividing a periodogram calculated for each segment by $U = \frac{1}{N_w}\sum_{n_w=0}^{N_w-1}\left(W(n_w)\right)^2$, where N_w is a total number of points in a window function, and n_w is an element counter in a window function with a range of $0 \le n_w \le N_w - 1$. Finally, a smoothed periodogram is obtained by averaging all periodograms calculated for these segments. The difference between the Bartlett's method and the Welch's method is that segments are overlapped and windowed in the Welch's method (Djuric et al., 1999; Gupta et al., 2013; Shradhanjali et al., 2013; Rahi and Mehra; 2014; Karuna and Mahalakshmi, 2015).

The main technique used in previous methods is to subdivide a data into several segments. Consequently, these subdividing of the data into a series of segments results in a lower frequency resolution and a higher frequency interval. To obtain a periodogram with finer frequency intervals, two methods can be used. One method is to calculate a smoothed periodogram in coarser frequency intervals and then interpolate it on a finer frequency resolution. In the second method, after data is subdivided into a series of segments, zero values are padded (added) to the end of each segment to make a number of elements in each segment equal to a number of elements in an original time series. Then, a periodogram is calculated for each of the zero-padded segment. As a result, a periodogram for each segment has the same frequency resolution as if an entire time series is used to calculate a periodogram. Finally, an averaged periodogram is obtained by averaging periodograms from all segments.

5.8 Calculating a power spectral density using MATLAB/Octave and Python/SciPy functions

Most of the programming languages have built-in functions to calculate a power spectral density using a periodogram or the Welch's methods. Now, a step by step practical guide is provided on how to use a built-in a periodogram and the Welch functions in MATLAB or GNU Octave. Then, a Python code is presented.

In practice, researchers work on time series signals that represent various types of situations, particularly the ones that are gathered from real phenomena. For example, a researcher might work on a water wave signal that is obtained from a measurement in the field or in a laboratory. Here, a simple synthetic time series is generated and used for the purpose of illustration in this example. Any other time series can be used as well. To generate this time series, a duration and a sampling frequency of the time series are given as:

```
%Input parameters
duration=1024; %Total duration of the time series,
duration=N*dt (here: duration=2^10)
fs=2; %Sampling frequency, fs=1/dt
```

Next, a time interval, Δt, and a time vector containing time values are calculated as:

```
%Generating a time vector from 0 to (duration-dt), equally
spaced at dt
dt=1/fs; %Time difference between consecutive samples,
dt=1/fs
t(:,1)=[0:dt:duration-dt]; %Time
```

Then, a simple water surface time series is generated by using a sinusoidal wave, i.e. $\eta(t) = a\cos(2\pi f t)$, with an amplitude of $a = 0.5\ m$ and a frequency of $f = 1/T = 0.2\ Hz$ as:

```
%Generating a water surface time series with an amplitude of
0.5 m and a frequency of 0.2 Hz (i.e. period of 5 s)
Eta(:,1)=0.5.*cos(2*pi*0.2*t);
```

After a sinusoidal signal is generated, a random signal, i.e. noise, is added to a sinusoidal signal as:

```
%Adding a random signal to a sinusoidal signal
rnd(:,1)=-0.1+(0.1-(-0.1)).*rand(length(t(:,1)),1);
%Uniformly distributed random signal between -0.1 m and 0.1
m
Eta=Eta+rnd; %Adding a random signal to a sinusoidal signal
```

For the purpose of the spectral analysis, the first step after a time series is obtained, is to remove any trend in the data or detrend the data. In other words, a mean value of the time series should be removed from a dataset in a way that a mean value of the time series becomes zero. This detrending must be done as the first step for almost any signal analysis calculation.

```
%Detrending the data
Eta=detrend(Eta);
```

Now, a frequency vector containing $\left(\frac{N}{2}+1\right)$ frequency values in a range of $0 \leq f \leq f_s/2$, where f_s is a sampling frequency, f_{Ny} is the Nyquist frequency ($f_s/2 = f_{Ny}$), and N is a total number of samples in a time series, is generated as:

```
%Generating a frequency vector
N=length(Eta(:,1)); %Length of the time series
df=fs/N; %Frequency difference between consecutive samples,
df=fs/N
f(:,1)=[0:df:fs/2]; %Frequency vector from 0 Hz to fNy=fs/2
Hz, equally spaced at df
```

Now, a periodogram of the water surface elevation is calculated as:

```
%Calculating a power spectral density (psd) using
MATLAB/Octave periodogram function
[psdEta_1Sided_PGram,fPGram]=periodogram(Eta,[],N,fs);
```

The other widely used method to estimate a power spectral density is the Welch's method. The Welch's method first divides data into a series of overlapping segments. It usually divides data into 8 segments with 50% overlapping. Then it computes a periodogram for each segment and then averages all periodograms. Using the same water surface generated above, a pwelch function in MATLAB or GNU Octave with a window size of 256 elements, a default Hamming window function, and 50% default overlapping is:

```
%Calculating a power spectral density (psd) using
MATLAB/Octave pwelch function
[psdEta_1Sided_Welch,fWelch]=pwelch(Eta,256,[],N,fs);
```

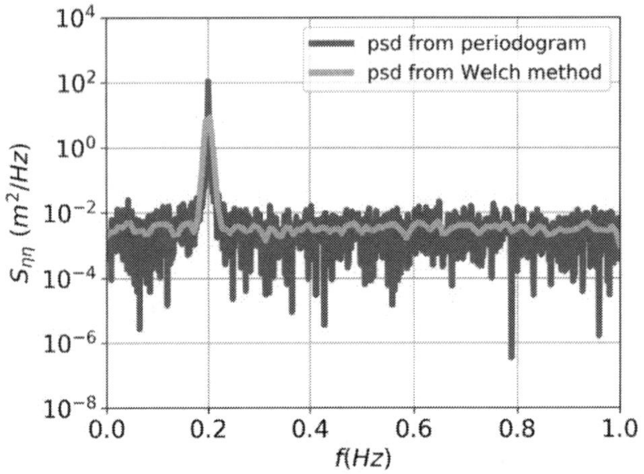

Figure 5.11 A power spectral density, *psd*, calculated by Python periodogram and Welch functions. An original time series has a primary frequency of $f = 0.2\ Hz$ and a sampling frequency of $f_s = 2\ Hz$. A spectrum has a frequency range of $0 \le f \le f_s/2$, i.e. $0 \le f \le 1$, with a peak wave frequency of $f_p = 0.2\ Hz$.

Now, results from periodogram and pwelch functions are plotted as:

```
%Plotting a power spectral density from MATLAB/Octave
functions
semilogy(fPGram,psdEta_1Sided_PGram)
hold on
semilogy(fWelch,psdEta_1Sided_Welch)
title('Power spectral density')
xlabel('Frequency (Hz)')
ylabel('Power/Frequency (m^2/Hz)')
legend('psd from periodogram','psd from Welch method')
```

Here is a Python code for calculating a power spectral density using periodogram and Welch's methods. Note that you need to install and import NumPy package as np, SciPy package as sp, and matplotlib.pyplot package as plt along with Python.

```
#-----------------------------------------------------------------
#Importing required packages

import numpy as np
import scipy as sp
from scipy import signal
import matplotlib.pyplot as plt

#-----------------------------------------------------------------

#Input parameters
duration=1024; #Total duration of the time series, duration=N*dt
(here: duration=2^10)
fs=2; #Sampling frequency, fs=1/dt

#-----------------------------------------------------------------
#Generating a signal

#Generating a time vector from 0 to (duration-dt), equally spaced
at dt
dt=1/fs; #Time difference between consecutive samples, dt=1/fs
t=np.arange(0,duration,dt); #Time

#Generating a water surface time series with an amplitude of 0.5
m and a frequency of 0.2 Hz (i.e. period of 5 s)
Eta=0.5*np.cos(2*np.pi*0.2*t);

#Adding a random signal to a sinusoidal signal
rnd=-0.1+(0.1-(-0.1))*np.random.rand(len(t)); #Uniformly
distributed random signal between -0.1 m and 0.1 m
Eta=Eta+rnd; #Adding a random signal to a sinusoidal signal

#Detrending the data
Eta=sp.signal.detrend(Eta)

#-----------------------------------------------------------------

#Generating a frequency vector
N=len(Eta); #Length of the time series
df=fs/N; #Frequency difference between consecutive samples,
df=fs/N
f=np.arange(0,fs/2+df,df); #Frequency vector from 0 Hz to
fNy=fs/2 Hz, equally spaced at df

#-----------------------------------------------------------------

#Calculating a power spectral density (psd) using Python
periodogram function
fPGram,psdEta_1Sided_PGram=sp.signal.periodogram(Eta,fs);

#Calculating a power spectral density (psd) using Python pwelch
function
```

```
fWelch,psdEta_1Sided_Welch=sp.signal.welch(Eta,fs,nperseg=256);

#-----------------------------------------------------------------

#Plotting a power spectral density from Python functions
plt.semilogy(fPGram,psdEta_1Sided_PGram,label='psd from
periodogram')
plt.semilogy(fWelch,psdEta_1Sided_Welch,label='psd from Welch
method')
plt.title('Power spectral density')
plt.xlabel('Frequency (Hz)')
plt.ylabel('Power/Frequency (m^2/Hz)')
plt.legend()

#-----------------------------------------------------------------
```

5.9 Code for smoothing a power spectral density using convolution

It is suggested to use the Welch's method to obtain a smoothed power spectral density. However, a simple method is presented here for smoothing a spectrum by using a convolution and the Hamming window function. Consider a periodogram of the water surface elevation that calculated above i.e. *psdEta_1Sided_PGram*. To smooth this periodogram, the Hamming window function is created and used. Here, a step by step practical guide is provided on how to smooth a periodogram by a window function using MATLAB or GNU Octave. Then, a Python code is presented. First, a normalized Hamming window with a window width of 11 elements is generated as:

```
%Creating a window function
Nw=11; %Number of elements in a window function
nwin(:,1)=[0:1:Nw-1]; %Counter from 0 to Nw-1
Wn=0.54-0.46*cos(2*pi*nwin/(Nw-1)); %Hamming window function
WnNorm=Wn./sum(Wn(:,1)); %Normalizing a window function
```

Now, a periodogram is smoothed by using a convolution as:

```
%Smoothing a power spectral density using a convolution
psdEta_1Sided_Smoothed=conv(psdEta_1Sided_PGram,WnNorm,'same
'); %Smoothing a power spectral density
```

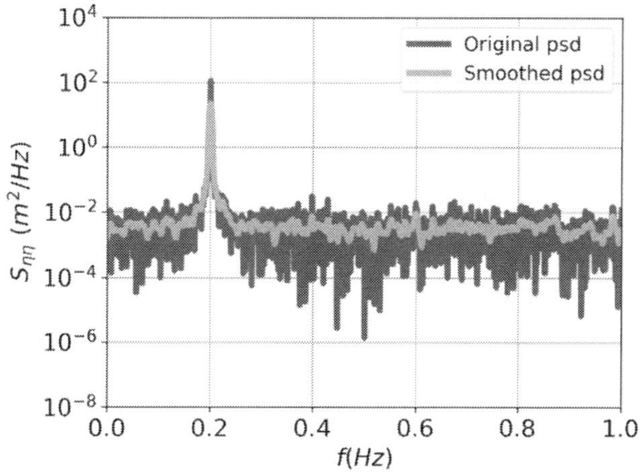

Figure 5.12 An original and smoothed power spectral densities, *psd*, using the Hamming window function containing 11 elements. An original time series has a primary frequency of $f = 0.2\ Hz$ and a sampling frequency of $f_s = 2\ Hz$. A spectrum has a frequency range of $0 \leq f \leq f_s/2$, i.e. $0 \leq f \leq 1$, with a peak wave frequency of $f_p = 0.2\ Hz$.

Note that the smoothing procedure should not change an area under a spectrum. In other words, $\int_0^\infty S_{\eta\eta}(f)df = \int_0^\infty S_{\eta\eta-Smooth}(f)df$. Now, both original and smoothed spectra are plotted as:

```
%Plotting a smoothed power spectral density
semilogy(f, psdEta_1Sided_PGram)
hold on
semilogy(f,psdEta_1Sided_Smoothed)
title('Power spectral density')
xlabel('Frequency (Hz)')
ylabel('Power/Frequency (m^2/Hz)')
legend('Original psd','Smoothed psd');
```

Here is a Python code for smoothing a periodogram. Note that you need to install and import NumPy package as np and matplotlib.pyplot package as plt along with Python.

```
#-----------------------------------------------------------------
#Importing required packages

import numpy as np
import matplotlib.pyplot as plt

#-----------------------------------------------------------------

#Creating a window function
Nw=11; #Number of elements in a window function
nwin=np.arange(0,Nw,1); #Counter from 0 to Nw-1
Wn=0.54-0.46*np.cos(2*np.pi*nwin/(Nw-1)); #Hamming window function
WnNorm=Wn/np.sum(Wn); #Normalizing a window function in time
domain

#-----------------------------------------------------------------
#Smoothing a power spectral density using a convolution

psdEta_1Sided_Smoothed=np.convolve(psdEta_1Sided_PGram,WnNorm,'sam
e'); #Smoothing a power spectral density

#-----------------------------------------------------------------

#Plotting a smoothed power spectral density
plt.semilogy(f, psdEta_1Sided_Smoothed,label='Original psd')
plt.semilogy(f,psdEta_1Sided_Smoothed,label='Smoothed psd')
plt.title('Power spectral density')
plt.xlabel('Frequency (Hz)')
plt.ylabel('Power/Frequency (m^2/Hz)')
plt.legend()

#-----------------------------------------------------------------
```

6 Pressure Data Analysis

In this chapter, calculation of the wave properties from measured pressure data is covered. First, a pressure data correction is explained. Next, estimation of the wave properties in the time and frequency domains is described. Then, a procedure for using a wave spectrum diagnostic tail is described. Step by step method to calculate wave properties from pressure data in the frequency domain is presented.

6.1 Pressure data correction to account for dynamic pressure attenuation in depth

In general, two types of instruments are widely used for a direct measurement of the wave activities. The first ones are instruments that collect water surface elevation data. The second ones are instruments that collect water pressure data associated with water surface elevation activities (i.e. water pressure associated with wave activities). If an instrument collects water surface elevation data, then data can be used right away for the wave analysis after passed through quality checks. However, if an instrument collects water pressure data associated with the wave activities, then pressure data need to be converted to water surface elevations before advancing toward the wave analysis. This conversion usually is not straightforward and needs extra attention and skills.

If collected data are water pressure associated with the wave activities, then recorded data consist of two main signals. The first signal shows hydrostatic pressure and the second signal shows dynamic pressure. The hydrostatic pressure is the still-water pressure at a pressure sensor's depth. Because of that, hydrostatic pressure is used to estimate a water depth at a deployment location. The dynamic pressure is the water pressure caused by wave motions. In other words, water surface fluctuations or wave activities on a water surface cause water pressure fluctuations under a water surface. Because of that, dynamic pressure is used to calculate wave properties.

For the purpose of data analysis, it is described that a time series is commonly split into a series of data blocks (data bursts) with a specific time interval. For example, data might split into a series of 1-hour bursts. Following the same procedure, it is required to split a pressure time series into a series of data bursts first. Then, the total measured pressure in each burst needs to be split into hydrostatic pressure and dynamic pressure. To separate these two, first, hydrostatic pressure is calculated by averaging pressure data over time. Then, dynamic pressure is obtained by subtracting the hydrostatic pressure from the total (original) pressure data. Therefore, hydrostatic pressure in each burst of data is equal to a mean value of the pressure data in that burst, and dynamic pressure in each burst of data is equal to de-trended pressure data in that burst. Remember that data in each burst should be stationary for this purpose.

After dynamic pressure data are obtained by de-trending the original pressure data, then they need to be prepared for the wave analysis. A reason that dynamic pressure cannot be used right away after measurement for the wave analysis is that dynamic pressure, which is resulted from water surface fluctuations, attenuates in the water column as depth increases (e.g. Cavaleri, 1980). In other words, if a pressure sensor is moving down from a water surface towards a seabed, it records smaller and smaller dynamic pressure values for the similar wave as it moves down. If a pressure sensor keeps moving down, eventually it reaches a depth where dynamic pressure is almost damped out, and because of that, a sensor cannot sense dynamic pressure anymore. In such conditions, a pressure sensor only detects and records hydrostatic pressure.

The consequence of the dynamic pressure attenuation in water depth is that the dynamic pressure recorded by a pressure transducer located under a water surface is smaller and weaker compared with actual values of the dynamic pressure right beneath a water surface. Using these weaker dynamic pressure data for wave analysis will lead to an underestimation of the wave height and energy. Therefore, dynamic pressure data collected by a pressure transducer need to be corrected for the pressure loss in water depth before being used for any data analysis.

To describe a procedure for dynamic pressure data correction, first, measured pressure data are written as a combination of the hydrostatic and dynamic pressures as:

$$P_T(z=-h_s) = P_0 + P = \rho g h_s + \rho g \eta K_p \tag{6.1}$$

where, P_T is the total measured pressure at $z=-h_s$, the $P_0 = \rho g h_s$ is the hydrostatic part of the measured pressure at $z=-h_s$ which is equal to mean water pressure, and $P = \rho g \eta K_p$ is the dynamic part of the measured pressure at $z=-h_s$ which is equal to pressure resulted from water surface fluctuations. The z is an upward vertical axis with zero at a water surface, h_s is a sensor depth which is equal to $z=-h_s$, the ρ is the water density, g is the gravitational acceleration, η is an original water surface elevation, and K_p is a pressure response factor which represents a dynamic pressure to water surface elevation conversion factor. A pressure response factor is defined as:

$$K_p(f, z=-h_s) = \frac{\cosh k(h+z)}{\cosh(kh)} = \frac{\cosh(kd_s)}{\cosh(kh)} \tag{6.2}$$

where, K_p is a pressure response factor which represents a dynamic pressure to surface elevation conversion factor, h is local water depth, d_s is a pressure sensor location distance from a seabed, $f = 1/T$ is a wave frequency, k is a wavenumber, and T is a wave period. A wavenumber is a function of the wave frequency, $f = 1/T$, and local water depth, h. A wavenumber is derived from a water wave dispersion relationship, i.e. $\omega^2 = (2\pi f)^2 = gk\tanh(kh)$, where $\omega = 2\pi f = 2\pi/T$ is an angular frequency.

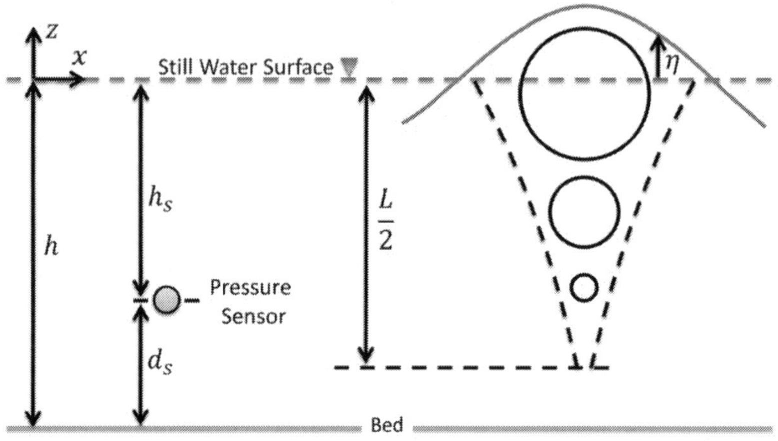

Figure 6.1 Schematic drawing of the water particles orbital motion and a pressure sensor deployment in deep water.

After a dynamic part of the measured pressure data is separated and a pressure response factor is calculated, dynamic pressure can be corrected for the pressure loss in water depth. To achieved this, dynamic pressure data are divided by a pressure response factor. Therefore, an original water surface elevation, η, which is accounted for the dynamic pressure loss, is obtained as:

$$\eta = \frac{1}{K_p} \times \frac{P}{\rho g} \tag{6.3}$$

where, η is a water surface elevation. Now, these water surface elevation data can be used for wave data analysis in the time or frequency domains.

In the frequency domain, a power spectral density of the water surface elevation can be obtained from an original water surface calculated above. Alternatively, a power spectral density of the water surface elevation can be calculated from a power spectral density of the measured dynamic pressure data as:

$$S_{\eta\eta} = \frac{1}{K_p^2} \times \frac{S_{PP}}{\rho^2 g^2} \tag{6.4}$$

where, $S_{\eta\eta}$ is a water surface elevation power spectral density, and S_{PP} is a dynamic pressure power spectral density.

However, in practice, calculating an original water surface elevation in the time domain and its spectrum in the frequency domain from the measured dynamic pressure is not a straightforward procedure and requires series of steps before the final results obtained. In following sections, these procedures will be discussed in details.

6.2 Obtaining water surface elevation from measured pressure data in the time domain

One method to obtain a water surface elevation from measured pressure data is to analyze the data in the time domain. To obtain an original water surface elevation from measured dynamic pressure data, a pressure response factor, K_p, is required to be calculated. The K_p is a function of the water depth, h, and the wavenumber, k. Therefore, to calculate K_p, water depth and a wavenumber should be defined first. Water depth is easily calculated from measured hydrostatic pressure as:

$$h = h_s + d_s = \frac{P_0}{\rho g} + d_s \tag{6.5}$$

where, hydrostatic pressure, P_0, is simply obtained by averaging the total measured pressure data, P_T, over time or over one burst of data.

To calculate a wavenumber, a wave period needs to be defined first. However, a calculation of the wave period is not as straightforward as a calculation of the water depth. Before calculating a wave period, first, it is considered that pressure attenuation in the water column only has an effect on wave energy (wave height), and has not any effect on a period of the dynamic pressure oscillation. In other words, a period of the dynamic pressure oscillation under a water surface is the same as a wave period of the water surface oscillation at a water surface. By considering this fact, to obtain a wave period, first, measured dynamic pressure data are converted to a water surface elevation without applying K_p as:

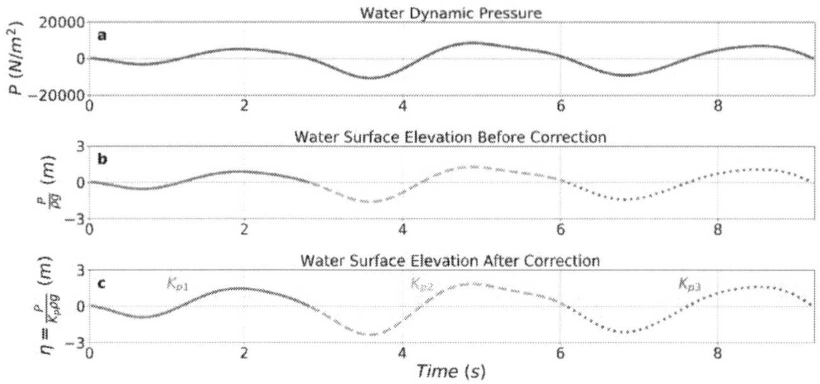

Figure 6.2 Converting water dynamic pressure to a water surface elevation. Plot (a) shows water dynamic pressure at a depth of 1 m below a water surface, i.e. $h_s = 1\ m$. Plot (b) shows the water dynamic pressure is divided by the water specific weight, i.e. ρg, and split into a series of individual waves by using a zero-crossing method. The ρ is the water density and g is the gravitational acceleration. In plot (c), every individual wave in a plot (b) is divided by an associated pressure response factor, K_p, to attain a water surface elevation, η. Here, a zero down-crossing is used, however, a zero up-crossing can be used as well. Synthetic pressure values in a plot (a) are obtained by generating a wave time series and multiplying it by the water specific weight and by a pressure response factor. A synthetic wave time series is resulted from a superposition of the second-order Stokes' waves with amplitudes, periods, and phases equal to $a = 0.5, 0.9, 0.2\ m, T = 4, 3, 2\ s$, and $\phi = 0.55\pi, -0.55\ \pi, 0.55\pi\ rad$, respectively. Water depth is equal to $h = 5\ m$. Note that only the first three waves are shown.

$$\eta_{Not\ Corrected} = \frac{P}{\rho g} \tag{6.6}$$

where, $\eta_{Not\ Corrected}$ is a water surface elevation obtained from dynamic pressure data without any correction for the pressure loss in the water column. Now, a time series of $\eta_{Not\ Corrected}$ is split into a series of single waves by using a zero-crossing method. Then, a wave period for every single wave in a time series is defined. From that, a wavenumber and a pressure response factor for each of those waves are calculated. After K_p is defined for each wave, $\eta_{Not\ Corrected}$ values corresponding to each of the split waves are converted to an original water surface elevation by using an associated K_p calculated for that wave as:

$$\eta = \frac{\eta_{Not\ Corrected}}{K_p} \tag{6.7}$$

After a water surface elevation corresponding to each of the split waves is calculated, an entire time series of the original water level is obtained by reassembling and putting the corrected water surface elevation associated with each split wave together. In other words, first, every single wave is corrected for the pressure loss, and then all of these waves are put back together in the same order that they are in a measured pressure dataset. Finally, wave properties are obtained from this corrected water surface elevation, η, by using a zero-crossing method.

Note that K_p has values between zero and one. As K_p approaches toward zero, $1/K_p$ approaches toward infinity. Therefore, to prevent an unacceptable magnification of the results as K_p approaches toward zero, a minimum value for a K_p should be considered, otherwise, it may result in very large values for very small K_p (for details, see pressure data analysis in the frequency domain).

6.3 Obtaining a power spectral density of the water surface elevation from measured pressure data in the frequency domain

It is more common to analyze wave data in the frequency domain rather than in the time domain. In other words, spectral analysis of wave data is a preferable method in most cases. One method to do that is to obtain a water surface elevation time series from pressure data in the time domain, and then use this surface elevation time series for the spectral analysis. However, it is not a usual practice as it is more common to directly calculate a wave spectrum from a pressure spectrum in the frequency domain. However, spectral analysis of the measured pressure data can be very challenging.

In spectral analysis, similar to the time domain procedure, a pressure response factor is used to convert pressure to a water surface elevation. However, this time, all calculations are carried out in the frequency domain. To do that, first, a pressure response factor is calculated for all frequencies in a range of $0 \leq f \leq f_s/2$, where f_s is a sampling frequency. Then, a power spectral density of the water surface elevation, $S_{\eta\eta}$, is calculated by using a power spectral density of the dynamic pressure, S_{PP}, as:

$$S_{\eta\eta}(f) = \frac{1}{\left(K_p(f)\right)^2} \times \frac{S_{PP}(f)}{\rho^2 g^2} \tag{6.8}$$

Then, wave properties are calculated from $S_{\eta\eta}$.

When data are analyzed in the frequency domain, it is necessary to remove all unreliable and noisy data from a spectrum. To remove these unreliable data, a low-cutoff frequency (a so-called high-pass filter), f_{cL}, or a high-cutoff frequency (a so-called low-pass filter), f_{cH}, or both f_{cL} and f_{cH} are used. In practice, f_{cL} and f_{cH} limits may be simply implemented by removing all data that have frequencies smaller than f_{cL}, i.e. $f < f_{cL}$, or have frequencies larger than f_{cH}, i.e. $f > f_{cH}$, from a dataset. In such cases, no transition band is considered between a passband and a stopband.

As noted above, when K_p approaches toward zero, $1/K_p$ approaches toward infinity. Therefore, similar to the time domain analysis, a minimum value for a K_p should be considered to prevent an unacceptable magnification of the results as K_p approaches toward zero, otherwise, it may result in very large values for very small K_p.

6.4 Lower limit for a pressure response factor

As mentioned above, in theory, a water surface elevation, η, time series in the time domain or a water surface elevation power spectral density, $S_{\eta\eta}$, in the frequency domain can be obtained by using a pressure response factor, K_p. However, in practice, an important question needs to be addressed first in order to get reliable values from these methods in the time and frequency domains. The question is: What is the smallest $K_p$value that can be used for a given dataset?

A reason that an answer to this question is crucial is that, in theory, a pressure response factor, K_p, which is a function of the wave frequency (wave period), has a range of $0 \le K_p \le 1$, where K_p is equal to $K_p = 1$ at $f = 0\ Hz$ and decreases as a frequency increases. This results in very small values for K_p in

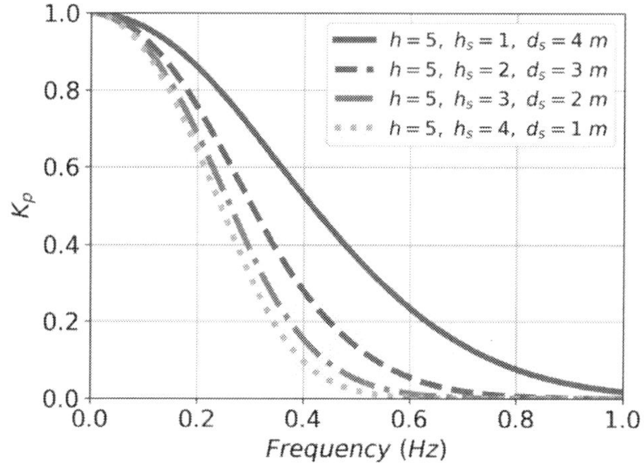

Figure 6.3 Schematic trend of the pressure response factor, K_p, versus a frequency. The K_p is calculated for a water depth equal to $h = 5\ m$ and four sensor depths equal to $h_s = 1, 2, 3, 4\ m$. Associated sensor heights from a seabed are $h_s = 4, 3, 2, 1\ m$, respectively. Note that only a frequency range of $0 \leq f \leq 1\ Hz$ is plotted.

higher frequencies. Eventually, as frequency increases further, K_p values approach zero, and as a result, $1/K_p$ values approach infinity. These small values for K_P (large values for $1/K_p$) lead to an unrealistic magnification of $S_{\eta\eta}$ values in higher frequencies. In other words, outcomes of $\eta = P/(\rho g K_p)$ in the time domain, and $S_{\eta\eta} = S_{PP}/(\rho g K_p)^2$ in the frequency domain become unrealistically large when K_p approaches zero for short waves. This unrealistic amplification of the short waves can lead to an error and unacceptable magnification in a calculation of the water level fluctuation or wave energy.

To prevent an error caused by small K_p values, a high-cutoff frequency, $f_{maxpcorr}$, which is associated with a minimum allowable pressure response factor, K_{pmin}, is chosen to limit a minimum value of the K_p to K_{pmin}. It means that, for frequency values larger than this cutoff frequency, i.e. for $f \geq f_{maxpcorr}$, values of K_p become smaller than K_{pmin}, and therefore, they should be limited to K_{pmin}. In other words, if $K_p < K_{pmin}$ then $K_p = K_{pmin}$. Limiting a minimum of the K_p value to K_{pmin} prevents amplification of the small waves

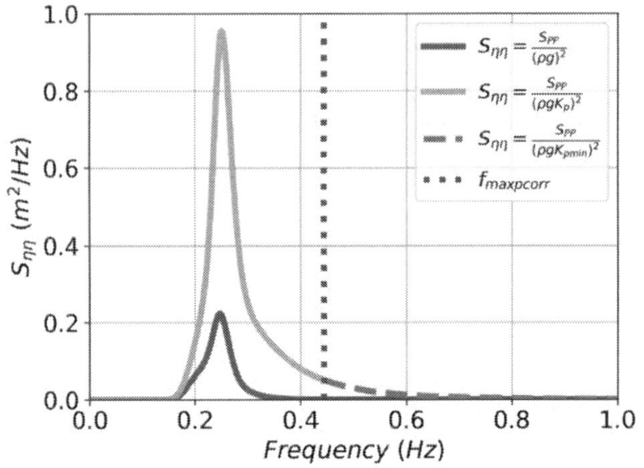

Figure 6.4 Schematic illustration of obtaining a water surface elevation power spectral density, $S_{\eta\eta}$, from a dynamic pressure power spectral density, S_{PP}. Plot shows $f_{maxpcorr}$ where a value of K_p is limited to K_{pmin} for a frequency range of $f \geq f_{maxpcorr}$. A schematic power spectral density is obtained from the JONSWAP spectrum (Hasselmann et al., 1973) for a wind velocity of 10 m/s, a wind fetch of 10000 m, and a peak wave frequency of 0.25 Hz.

or even noise with high frequencies. Therefore, this helps to avoid an error due to an over-estimation of the water surface oscillation or wave energy.

6.5 Theoretical method to define an acceptable minimum value for a pressure response factor

It is shown that for any given dataset, a minimum pressure response factor, K_{pmin}, and its associated frequency, $f_{maxpcorr}$, need to be defined. To define K_{pmin}, first, the smallest wave that its effect reaches the pressure sensor deployment depth should be defined. To find this smallest wave, the linear wave theory can be used.

The linear wave theory indicates that for any given wave, water particle motions resulted from water surface wave activities dissipate as water depth increases, and they nearly diminish in water depth larger than half of the wavelength. In other words, water surface fluctuation effects (wave effects) are only detectable from a water surface down to a depth equal to half of the wavelength of the surface wave, i.e. where $h \leq L/2$, where L is a wavelength.

Therefore, following the linear wave theory, a pressure sensor located at a depth equal to h_s can only detect waves with wavelengths that follow $L/2 \geq h_s$, and cannot sense waves with wavelengths that follow $L/2 \leq h_s$. It means that the smallest wave that is detectable by a pressure sensor located at a depth h_s, has a wavelength that satisfies $L/2 = h_s$. Because of that, all data with a wavelength that follows $L/2 \leq h_s$ are likely noise and should be removed from a dataset. Therefore, it is essential to define K_{pmin} and $f_{maxpcorr}$ accurately to ensure K_p is only being applied to waves that are sensed by a pressure sensor. This helps to prevent any amplification of the noise or waves that their effects do not reach a sensor's location.

Theoretically, values of K_{pmin} and $f_{maxpcorr}$ can be estimated by using the linear wave theory. As mentioned above, based on the linear wave theory, the water particle motion resulted from the wave oscillation is almost diminished at a depth equal to half of the wavelength of the surface wave. Therefore, a wavelength of the shortest linear wave, $L_{min-Linear}$, that is detectable by a pressure sensor located at depth h_s follows:

$$h_s = \frac{L_{min-Linear}}{2} \tag{6.9}$$

where, L is a wavelength, $L_{min-Linear}$ is a wavelength of the shortest linear wave that is detectable by a pressure sensor, h_s is a pressure sensor location depth, and a subscript *Linear* denotes that the linear wave theory is used to estimate a parameter. Considering a wavenumber is $k = 2\pi/L$, this relationship can be written as:

$$h_s = \frac{\pi}{k_{max-Linear}} \tag{6.10}$$

or

$$k_{max-Linear} h_s = \pi \tag{6.11}$$

where, $k_{max-Linear}$ is a wavenumber of the shortest linear wave that is detectable by a pressure sensor and is associated with $L_{min-Linear}$.

Considering $h_s = h - d_s$, where d_s is a pressure sensor distance from a seabed, this relationship can be re-written as:

$$k_{max-Linear}(h - d_s) = \pi \tag{6.12}$$

or

$$k_{max-Linear} = \frac{\pi}{h-d_s} \tag{6.13}$$

Then, a minimum acceptable value for a pressure response factor, estimated from the linear wave theory, $K_{pmin-Linear}$, is equal to:

$$K_{pmin-Linear}(f_{maxpcorr-Linear}, z = -h_s) = \frac{\cosh(k_{max-Linear} d_s)}{\cosh(k_{max-Linear} h)} \tag{6.14}$$

or

$$K_{pmin-Linear}(f_{maxpcorr-Linear}, z = -h_s) = \cosh\left(\frac{\pi}{h-d_s} d_s\right) / \cosh\left(\frac{\pi}{h-d_s} h\right) \tag{6.15}$$

To estimate a frequency associated with $K_{pmin-Linear}$, a water wave dispersion relationship is written as:

$$\omega^2 = (2\pi f)^2 = gk\tanh(kh) \tag{6.16}$$

where, $\omega = 2\pi f$ is a wave angular frequency, and k is a wavenumber. A dispersion relationship can be re-written as:

$$f = \frac{1}{2\pi}\sqrt{gk\tanh(kh)} \tag{6.17}$$

By replacing k with $k_{max-Linear}$, a frequency associated with $K_{pmin-Linear}$ is:

$$f_{maxpcorr-Linear} = \frac{1}{2\pi}\sqrt{g k_{max-Linear} \tanh(k_{max-Linear} h)} \tag{6.18}$$

or

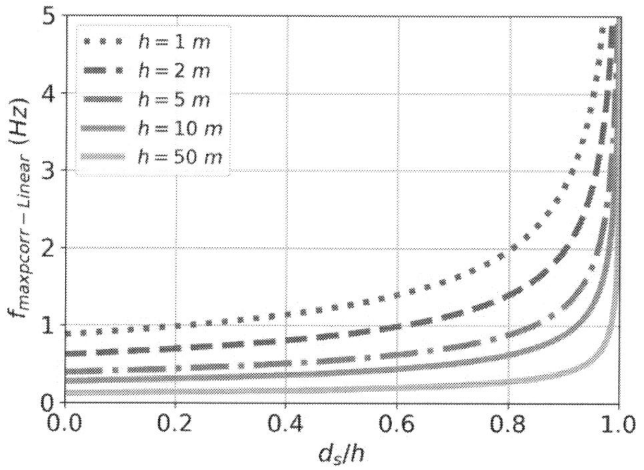

Figure 6.5 Trend of $f_{maxpcorr-Linear}$ versus d_s/h for water depth equal to $h = 1, 2, 5, 10, 50\ m$. The $f_{maxpcorr-Linear}$ shows a maximum frequency calculated from the linear wave theory beyond which $K_p = K_{pmin-Linear}$.

$$f_{maxpcorr-Linear} = \frac{1}{2\pi}\sqrt{g\frac{\pi}{h-d_s}\tanh\left(\frac{\pi}{h-d_s}h\right)} \tag{6.19}$$

where, $f_{maxpcorr-Linear}$ is a frequency associated with $K_{pmin-Linear}$. The $f_{maxpcorr-Linear}$ shows a maximum frequency estimated by the linear wave theory where at that frequency $K_p = K_{pmin-Linear}$. Therefore, for frequencies larger than $f_{maxpcorr-Linear}$, i.e. for $f \geq f_{maxpcorr-Linear}$, values of K_p become smaller than $K_{pmin-Linear}$, and because of that, values of K_p should be limited to $K_{pmin-Linear}$ in this frequency range.

If a pressure sensor sits on a seabed, then $d_s = 0$ and $z = -h = -h_s$. Therefore:

$$k_{max-Linear}h_s = k_{max-Linear}h = \pi \tag{6.20}$$

Then, $K_{pmin-Linear}$ for a pressure sensor that sits on a seabed is:

$$K_{pmin-Linear}\left(f_{maxpcorr-Linear}, z = -h\right) = \frac{\cosh(0)}{\cosh(\pi)} \approx 0.0862 \tag{6.21}$$

And, $f_{maxpcorr-Linear}$ for a pressure sensor that sits on a seabed is:

$$f_{maxpcorr-Linear} = \frac{1}{2\pi}\sqrt{g\frac{\pi}{h}\tanh(\pi)} \approx 0.8819h^{-0.5} \tag{6.22}$$

6.6 Practical methods to define an acceptable minimum value for a pressure response factor

It is explained how to use the linear wave theory to predict $f_{maxpcorr-Linear}$ and $K_{pmin-Linear}$ values. However, these theoretical values are only applicable to sinusoidal waves. It means that if measured dynamic pressure data are following a sinusoidal profile, then considering $f_{maxpcorr} \approx f_{maxpcorr-Linear}$ and $K_{pmin} \approx K_{pmin-Linear}$ leads to satisfactory results.

However, if a wave profile departs from a sinusoidal form then using $f_{maxpcorr-Linear}$ and $K_{pmin-Linear}$ in the data analysis is not acceptable anymore. Such conditions may occur when waves propagate into an intermediate and shallow water depth. Under these conditions, a waveform becomes skewed and does not follow a sinusoidal form anymore. When a waveform deviates from a sinusoidal form, then $f_{maxpcorr} < f_{maxpcorr-Linear}$ and $K_{pmin} > K_{pmin-Linear}$. It means that considering $f_{maxpcorr} \approx f_{maxpcorr-Linear}$ and $K_{pmin} \approx K_{pmin-Linear}$ for non-sinusoidal waves result in an over-estimation of the water surface oscillation and wave energy.

In practice, $f_{maxpcorr}$ and K_{pmin} need to be defined as accurately as possible to avoid signal inflation where $f > f_{maxpcorr}$. To do that, different techniques can be implemented in the time and frequency domains.

6.6.1 Practical method to define $f_{maxpcorr}$ and K_{pmin} in the time domain

To use K_{pmin} in the time domain data analysis, first, a constant predefined value for K_{pmin} is chosen. This would be the lowest value K_p can get in the analysis. For example, a constant predefined $K_{pmin} = 0.15$ is a justifiable choice. Considering a $K_{pmin} = 0.15$ helps to prevent unacceptable amplification of the measured data greater than 6.66 times (i.e. $1/0.15$ times).

Note that, although $K_{pmin} = 0.15$ might be suitable for most cases, this value needs to be chosen based on waves and water body conditions.

After this K_{pmin} value is set, all waves in a time series are split into a series of individual waves. Then, $K_{pmin-Linear}$ is calculated for each of these individual waves. Finally, a calculated $K_{pmin-Linear}$ for each wave is compared with a constant predefined K_{pmin}, and a maximum value between them, i.e. $max(K_{pmin-Linear}, K_{pmin})$, is used to convert pressure data to water surface elevation for that wave. This procedure is repeated for all individual waves in a time series.

6.6.2 Practical methods to define $f_{maxpcorr}$ and K_{pmin} in the frequency domain

In the frequency domain data analysis, there are three approaches that can be implemented to define $f_{maxpcorr}$ and K_{pmin}, which helps to avoid signal inflation where $f > f_{maxpcorr}$.

In the first approach, first, a water surface elevation power spectral density is calculated without applying K_p as $S_{\eta\eta} = S_{PP}/(\rho^2 g^2)$. Then, a noise floor at the end of the power spectral density is defined. Next, this noise floor is uniformly subtracted from a water surface elevation power spectral density before applying K_p. Finally, a water surface elevation power spectral density is recalculated by applying a K_p on the spectrum as $S_{\eta\eta} = (1/K_p^2) \times S_{PP}/(\rho^2 g^2)$ with its noise floor removed. In this approach, it is assumed that a noise floor is a uniformly distributed wide band noise for all frequencies (Bishop and Donelan, 1987; Trowbridge and Elgar, 2001; Smith, 2002; Jones and Monismith, 2007). This approach reflects an assumption that a frequency where a noise floor starts at a tail of the power spectral density is equal to $f_{maxpcorr}$. In other words, a frequency that a noise floor emerges at a tail of the power spectral density is a frequency of the shortest wave that a pressure sensor can detect. Therefore, effects of the shorter waves with larger wave frequencies cannot be detected by a pressure transducer as they diminished before reaching a sensor depth. Because of that, all the readings appear after $f_{maxpcorr}$ are considered as noise. Note that, a non-uniform noise floor subtraction from the spectrum is not recommended as it can lead to an under-estimation of the results (Jones and Monismith, 2007).

In the second approach, which is more common than the first one, first, a high-cutoff frequency equal to f_{cH} is selected and assumed it is equal to $f_{maxpcorr}$, i.e. it is assumed that $f_{maxpcorr} = f_{cH}$. Then, all data with frequencies $f > f_{cH}$ are filtered out of the power spectral density. The f_{cH} is chosen in a way to assure removal of all data with a low signal-to-noise ratio (Bishop and Donelan, 1987; Smith, 2002; Jones and Monismith, 2007). A simple method to define f_{cH} is to set f_{cH} equal to a frequency that a noise floor starts at the tail of the power spectral density. In this method, a high-cutoff frequency can be constant (the same) for all bursts in a dataset, or it can be selected adaptively for each burst. Between these two, a selection of the constant cutoff frequency, i.e. $f_{maxpcorr} = f_{cH} =$ Constant, for all bursts is recommended. However, selecting a constant cutoff frequency for all bursts is not a straightforward procedure and requires a series of criteria to be met (Jones and Monismith, 2007), and still may lead to an inaccurate wave calculation (e.g. Van Rijn et al., 2000). Note that, selecting an adaptive cutoff frequency, i.e. $f_{maxpcorr} = f_{cH} =$ Adaptive, for each burst is not suggested as it can result in dissimilar conditions from burst to burst.

The third approach is a modified form of the second approach, aimed to improve the second approach for defining $f_{maxpcorr}$ and K_{pmin}. In this approach, two different values are chosen for f_{cH} and $f_{maxpcorr}$, where $f_{maxpcorr} \leq f_{cH}$. In most cases, it is an acceptable practice in the frequency domain data analysis to use a constant high-cutoff frequency, i.e. $f_{cH} =$ Constant, for all bursts. In contrast, values of $f_{maxpcorr}$ and K_{pmin} can be either constant for all bursts or can be selected adaptively for each burst. However, in most situations, it is inadequate to consider a constant value for $f_{maxpcorr}$ and K_{pmin}, because it can result in an incorrect wave estimation due to either over or under-estimation of the wave energy. Therefore, in the third approach, a constant value is considered for f_{cH} while adaptive values are implemented for $f_{maxpcorr}$ and K_{pmin}, unless using a constant value for $f_{maxpcorr}$ and K_{pmin} is justifiable. In other words, while $f_{cH} =$ Constant for the entire dataset, values of $f_{maxpcorr}$ and K_{pmin} either are $f_{maxpcorr} =$ Adaptive and $K_{pmin} =$ Adaptive, which are defined for each burst individually, or are $f_{maxpcorr} =$ Constant and $K_{pmin} =$ Constant for the entire dataset. A constant value for f_{cH} guarantees a similar high-cutoff frequency for all wave spectrums in case that a series of wave spectrums are calculated for a series of bursts. Calculating the adaptive values for $f_{maxpcorr}$ and K_{pmin} in each burst help to avoid noise inflation, as a noise floor start-frequency may differ from burst to burst. Then, unlike the second approach that assumes $f_{maxpcorr} = f_{cH}$, in the third approach $f_{maxpcorr} \leq f_{cH}$.

6.6.3 Some guidelines on defining a constant f_{cH}, an adaptive $f_{maxpcorr}$, and an adaptive K_{pmin} in the frequency domain

If the third approach is selected for data analysis, it is recommended to use a constant value for f_{cH}, and adaptive values for $f_{maxpcorr}$ and K_{pmin}. To use this method, these values, i.e. a constant value for f_{cH}, and adaptive values for $f_{maxpcorr}$ and K_{pmin}, need to be defined. In this approach, if only one spectrum needs to be evaluated, then defining f_{cH} = Constant, $f_{maxpcorr}$ = Adaptive and K_{pmin} = Adaptive might not be very complicated. However, in most studies, a series of data bursts are collected, and a power spectral density needs to be calculated for each individual burst. In such conditions, which a large number of spectrums should be calculated, defining f_{cH} = Constant that works for all spectrums and estimating $f_{maxpcorr}$ = Adaptive and K_{pmin} = Adaptive for each individual spectrum can be very challenging.

Selecting a constant f_{cH} value that suits all spectrums involves in finding a value for f_{cH} which eliminates noise at a tail of all power spectral densities. This is commonly done by visually inspecting a series of spectrums before applying K_p, i.e. it is done by inspecting $S_{\eta\eta} = S_{PP}/(\rho^2 g^2)$. This visual inspection helps to identify a single value for f_{cH} which filters out noise in all spectrums.

Defining adaptive values of $f_{maxpcorr}$ and K_{pmin} for each burst is not as straightforward. In general, a value of $f_{maxpcorr}$ for each burst of data is defined by searching for an exact location of $f_{maxpcorr}$ in a power spectral density of that burst. There are two methods to search and find the values of $f_{maxpcorr}$ and K_{pmin} in each burst.

In the first method, first, a water surface elevation power spectral density is calculated without applying K_p as $S_{\eta\eta} = S_{PP}/(\rho^2 g^2)$. Next, a noise floor at the end of the power spectral density is identified. Then, $f_{maxpcorr}$ is set equal to a frequency that a noise floor emerges at a tail of the power spectral density.

It is based on the assumption that $f_{maxpcorr}$ represents a frequency of the shortest wave that a pressure sensor can detect, and typically it is equal to a frequency where a noise floor starts at a tail of the power spectral density.

In the second method, first, a water surface elevation power spectral density is calculated by applying K_p without limiting K_p to K_{pmin} as $S_{\eta\eta} = (1/K_p^2) \times S_{PP}/(\rho^2 g^2)$, where K_p has a range of $0 \leq K_p \leq 1$. By not limiting K_p to K_{pmin}, a calculated spectrum at first increases to its peak value and then it starts to decrease until it reaches to a minimum value. After dropping to this minimum point, it starts to rise again and moves toward infinity. A frequency of this minimum value in a spectrum tail before it moves toward infinity is $f_{maxpcorr}$. Note that a location of this minimum is after a spectrum peak frequency, f_p, in a tail of the spectrum.

After a value of $f_{maxpcorr}$ is defined by either of these methods, a minimum value of the pressure response factor, K_{pmin}, which is a K_p value associated with $f_{maxpcorr}$, is calculated. Then, a pressure response factor is calculated for all frequencies in a range of $0 \leq f \leq f_s/2$. After that, the lower boundary of K_p is limited to K_{pmin} as $K_{pmin} \leq K_p \leq 1$, where $K_p = 1$ at $f = 0\ Hz$ and decreases to $K_p = K_{pmin}$ at $f = f_{maxpcorr}\ Hz$.

Note that in this approach, the adaptive $f_{maxpcorr}$ and K_{pmin} should always be limited to values obtained from the linear wave theory as $f_{maxpcorr} \leq f_{maxpcorr-Linear}$ and $K_{pmin} \geq K_{pmin-Linear}$.

If higher harmonics of the dominant frequency have considerable energies, then a water surface elevation power spectral density might have multiple energy peaks, which are associated with those higher harmonics. In particular, this may be a case when the 2nd harmonic is presented in the data. In such cases, additional attention requires to avoid identifying these higher harmonics mistakenly as a high-frequency amplification caused by the small K_p values. To do that, an adaptive $f_{maxpcorr}$ needs to be chosen in a way that it reflects the higher harmonic effects without amplifying the noise.

After $f_{maxpcorr}$ is defined, then it is compared with f_{cH}. If a value of $f_{maxpcorr}$ becomes equal or larger than a constant value chosen for f_{cH}, i.e. if $f_{maxpcorr} \geq f_{cH}$, then a value of $f_{maxpcorr}$ is limited to f_{cH} as $f_{maxpcorr} = f_{cH}$. Otherwise, if

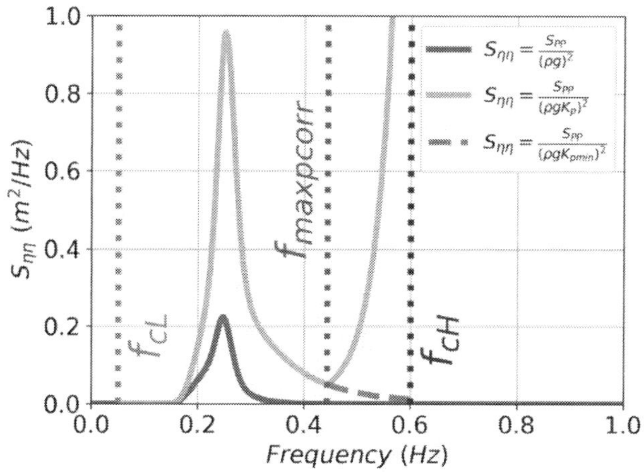

Figure 6.6 Schematic illustration of obtaining a water surface elevation power spectral density, $S_{\eta\eta}$, from a dynamic pressure power spectral density, S_{PP}. Plot shows how high-frequency energy values are amplified if values of K_p are not limited by K_{pmin}. To prevent artificial amplification of the spectrum, $f_{maxpcorr}$ is chosen to limits K_p to K_{pmin} for a frequency range of $f \geq f_{maxpcorr}$. A low cut-off frequency is $f_{cL} = 0.05\ Hz$, a high cut-off frequency is $f_{cH} = 0.6\ Hz$, and a sampling frequency is $f_s = 2\ Hz$. Values of f_{cL}, f_{cH} and f_s in the figure are selected for a demonstration purpose. In practice, these values need to be selected based on the physical properties of the study area, instrumentation, and nature of the research. A schematic power spectral density is obtained from the JONSWAP spectrum (Hasselmann et al., 1973) for a wind velocity of 10 m/s, a wind fetch of 10000 m and a peak wave frequency of 0.25 Hz.

a value of $f_{maxpcorr}$ is smaller than f_{cH}, i.e. if $f_{maxpcorr} < f_{cH}$, then additional attention is required for applying K_p within a frequency range of $f_{maxpcorr} \leq f \leq f_{cH}$.

If $f_{maxpcorr} < f_{cH}$, then two approaches may be considered to apply K_p to the data within a frequency range of $f_{maxpcorr} \leq f \leq f_{cH}$. In the first approach, a transition band is considered from $f_{maxpcorr} - \Delta f_{tr}$ to $f_{maxpcorr} + \Delta f_{tr}$, where $\pm\Delta f_{tr}$ defines a frequency range for a transition band before and after $f_{maxpcorr}$ and is chosen based on the spectrum's conditions. Then, K_p is gradually increased from $K_p = K_{pmin}$ at $f = f_{maxpcorr} - \Delta f_{tr}$ to $K_P = 1$ at $f =$

$f_{maxpcorr} + \Delta f_{tr}$ and retained equal to 1 for higher frequencies, i.e. $K_p = 1$ for $f_{maxpcorr} + \Delta f_{tr} \leq f \leq f_{cH}$.

In the second approach, a value of K_P is kept equal to K_{pmin} for frequencies larger than $f_{maxpcorr}$, i.e. $K_p = K_{pmin}$ for $f_{maxpcorr} \leq f \leq f_{cH}$ (Huang and Tsai, 2008). A difference between these two approaches is that, in the second approach, where $K_p = K_{pmin}$ for $f_{maxpcorr} \leq f \leq f_{cH}$, a slope of the spectrum tail obtained from $S_{\eta\eta} = (1/K_p^2) \times S_{PP}/(\rho^2 g^2)$ is similar to a slope of the spectrum tail obtained from $S_{\eta\eta} = S_{PP}/(\rho^2 g^2)$. However, in the first approach, an increase of K_p from $K_p = K_{pmin}$ to $K_p = 1$ leads to dissimilar spectrum tail slopes for $S_{\eta\eta} = (1/K_p^2) \times S_{PP}/(\rho^2 g^2)$ compared to $S_{\eta\eta} = S_{PP}/(\rho^2 g^2)$.

Finally, after K_{pmin} is defined and K_p is calculated for all frequencies, a water surface elevation power spectral density, $S_{\eta\eta}$, is calculated from $S_{\eta\eta} = (1/K_p^2) \times S_{PP}/(\rho^2 g^2)$, where $K_{pmin} \leq K_p \leq 1$.

6.6.4 Step by step procedure to obtain a water surface elevation power spectral density from measured pressure data in the frequency domain

All aforementioned steps are required to be followed to attain accurate results from the spectral analysis. However, following these steps can be very overwhelming, particularly for defining adaptive values of $f_{maxpcorr}$ and K_{pmin}. To provide a better guideline, a step by step procedure, which is applicable in most cases, including how to define and implement adaptive values of $f_{maxpcorr}$ and K_{pmin}, is described. To convert measured pressure data to water surface elevation in the frequency domain, it is recommended to follow these steps:

- Calculate a water surface elevation power spectral density without using a pressure response factor as $S_{\eta\eta} = S_{PP}/(\rho^2 g^2)$.
- Visually evaluate a series of wave spectrums (as many as needed) and define a minimum frequency where spectrum values with smaller frequencies are not reliable. Set this frequency equal to f_{cL}. This indicates that spectrum values with a frequency range of $f < f_{cL}$ are not reliable. As this value of f_{cL} is used for all spectra, it might be

required to evaluate a series of spectra to find a f_{cL} value that is applicable for all spectra that are intended to be analyzed.

- Visually evaluate a series of wave spectrums (as many as needed) and define a maximum frequency which is required for the data analysis. In other words, decide about the highest frequency limit, where the spectrum values should be retained up to that point in order to carry out this particular research. Set this frequency equal to f_{cH}. There might be data with a frequency range of $f_{maxpcorr} \leq f \leq f_{cH}$ or $f_{tail} \leq f \leq f_{cH}$ where spectrum data are not reliable within that range. In such cases, either a value of f_{cH} should be decreased to a frequency that prevents these unreliable values retained in a spectrum, or these data need to be replaced by an empirical spectrum diagnostic tail. As this value of f_{cH} is used for all spectra, it might be required to evaluate a series of spectra to find a value for f_{cH} that is applicable for all spectra that are intended to be analyzed.
- Visually evaluate a series of wave spectrums (as many as needed) and define a frequency that a noise floor starts at a high-frequency range of the spectrum. A frequency that a noise floor starts is equal to $f_{NoiseFloor}$. Spectrum values with a frequency range of $f > f_{NoiseFloor}$ are not reliable. As this value of $f_{NoiseFloor}$ is used for all spectra, it might be required to evaluate a series of spectra to find a $f_{NoiseFloor}$ value that is applicable for all spectra that are intended to be analyzed.
- Define a peak wave frequency, f_p, of the spectrum. It is a frequency which is associated with a maximum value of the spectrum.
- Calculate a pressure response factor, K_p, for a frequency range of $0 \leq f \leq f_s/2$. A value of K_p is equal $K_p = 1$ at $f = 0\,Hz$ and starts to decrease toward zero as a frequency increases. A minimum value of K_p is obtained for $f = f_s/2$. If a sampling frequency, f_s, becomes large enough, then K_p approaches zero.
- Calculate a water surface elevation power spectral density by applying K_p without limiting K_p to K_{pmin} as $S_{\eta\eta} = \left(1/K_p^2\right) \times S_{PP}/(\rho^2 g^2)$ for a frequency range of $0 \leq f \leq f_s/2$. The K_p values can have a range of $0 \leq K_p \leq 1$ depending on a range of $0 \leq f \leq f_s/2$.
- Remove all values of the spectrum that have a frequency range of $f < f_{cL}$ and $f > f_{cH}$. In other words, only retain spectrum values with a frequency range of $f_{cL} \leq f \leq f_{cH}$.

- Find a minimum value of the spectrum in a frequency range of $f_p \leq f \leq f_{cH}$. Set this frequency equal to $f_{MinPoint}$.
- Set $f_{maxpcorr}$ equal to a smaller value between a frequency associated with a minimum point after peak frequency, i.e. $f_{MinPoint}$, and a frequency that a noise floor starts, i.e. $f_{NoiseFloor}$. In other words, set $f_{maxpcorr} = min(f_{MinPoint}, f_{NoiseFloor})$.
- From the linear wave theory, calculate $f_{maxpcorr-Linear}$ and compare it with $f_{maxpcorr}$. If $f_{maxpcorr}$ is larger than $f_{maxpcorr-Linear}$, then set $f_{maxpcorr}$ equal to $f_{maxpcorr-Linear}$. In other words, if $f_{maxpcorr} > f_{maxpcorr-Linear}$ then $f_{maxpcorr} = f_{maxpcorr-Linear}$. Remember that $f_{maxpcorr} \leq f_{maxpcorr-Linear}$ should be satisfied for all cases.
- Calculate a pressure response factor, K_p, associated with $f_{maxpcorr}$ and set it equal to K_{pmin}.
- Calculate a water surface elevation power spectral density by applying K_p, but this time limit a K_p value to K_{pmin} as $S_{\eta\eta} = (1/K_p^2) \times S_{PP}/(\rho^2 g^2)$, where $K_{pmin} \leq K_p \leq 1$ for a frequency range of $0 \leq f \leq f_s/2$.
- Remove all values of the spectrum that have a frequency range of $f < f_{cL}$ and $f > f_{cH}$. In other words, only retain spectrum values with a frequency range of $f_{cL} \leq f \leq f_{cH}$.
- If required, replace values of the spectrum tail with a frequency range of $f_{maxpcorr} \leq f \leq f_{cH}$ or $f_{tail} \leq f \leq f_{cH}$ with an empirical wave spectrum diagnostic tail (see wave spectrum diagnostic tail for detail).

Note that if constant values for $f_{maxpcorr}$ and K_{pmin} are chosen, then the same procedure with excluding the steps that define adaptive values for $f_{maxpcorr}$ and K_{pmin} can be followed.

Although this step by step guideline is applicable in most cases, note that there might be cases that this guideline does not produce reliable results. Therefore, in any data analysis, outcomes should be evaluated accurately to assure the quality of final results.

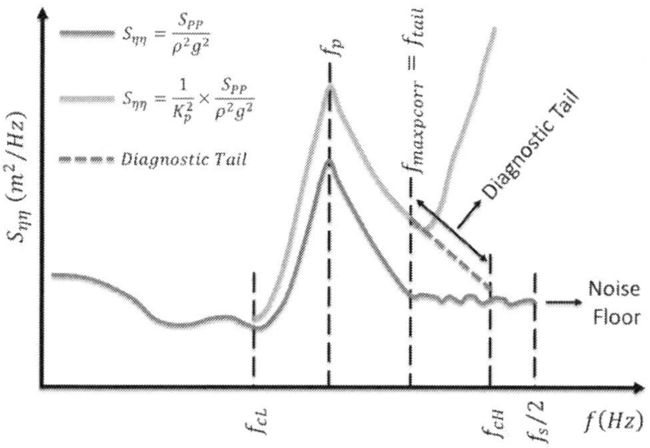

Figure 6.7 Schematic illustration of defining a low-cutoff frequency, f_{cL}, a high-cutoff frequency, f_{cH}, and $f_{maxpcorr}$. Plot shows a schematic water surface elevation power spectral density without using a pressure response factor as $S_{\eta\eta} = S_{PP}/(\rho^2 g^2)$, within a frequency range of $0 \leq f \leq f_s/2$, and with using a pressure response factor as $S_{\eta\eta} = (1/K_p^2) \times S_{PP}/(\rho^2 g^2)$, within a frequency range of $f_{cL} \leq f \leq f_{cH}$. In the plot, all values with a frequency range of $f_{cL} \leq f \leq f_{cH}$ are retained, and all values with a frequency range of $f < f_{cL}$ or $f > f_{cH}$ are removed. The $f_{maxpcorr}$ is selected equal to a frequency that a noise floor starts at a tail of the spectrum. The f_{tail} is chosen equal to $f_{maxpcorr}$. A spectrum diagnostic tail is used for a frequency range of $f_{maxpcorr} \leq f \leq f_{cH}$. All frequency values and a diagnostic tail in the figure are for a demonstration purpose. In practice, these values need to be selected based on the physical properties of the study area, instrumentation, and nature of the research. Similarly, a diagnostic tail might or might not be required.

6.7 Accuracy of using a pressure response factor to convert pressure data to water surface elevation

A described method above, which use a pressure response factor to convert pressure data to water surface elevation, is based on the linear wave theory. However, there are many instances in the real world that a wave profile does not follow the linear wave theory. A deviation of wave profile from the linear wave theory may raise concern about how accurate is the outcomes when a pressure response factor based on the linear wave theory is implemented to convert pressure data to water surface elevation.

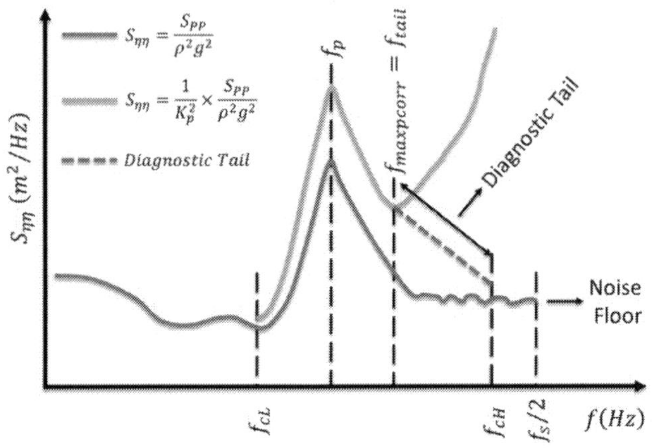

Figure 6.8 Schematic illustration of defining a low-cutoff frequency, f_{cL}, a high-cutoff frequency, f_{cH}, and $f_{maxpcorr}$. Plot shows a schematic water surface elevation power spectral density without using a pressure response factor as $S_{\eta\eta} = S_{PP}/(\rho^2 g^2)$, within a frequency range of $0 \leq f \leq f_s/2$, and with using a pressure response factor as $S_{\eta\eta} = \left(1/K_p^2\right) \times S_{PP}/(\rho^2 g^2)$ within a frequency range of $f_{cL} \leq f \leq f_{cH}$. In the plot, all values with a frequency range of $f_{cL} \leq f \leq f_{cH}$ are retained, and all values with a frequency range of $f < f_{cL}$ or $f > f_{cH}$ are removed. The $f_{maxpcorr}$ is selected equal to a frequency where spectrum tail starts to rise to infinity. The f_{tail} is chosen equal to $f_{maxpcorr}$. A spectrum diagnostic tail is used for a frequency range of $f_{maxpcorr} \leq f \leq f_{cH}$. All frequency values and a diagnostic tail in the figure are for demonstration purpose. In practice, these values need to be selected based on the physical properties of the study area, instrumentation, and nature of the research. Similarly, a diagnostic tail might or might not be required.

Some studies attempted to address this concern by suggesting an implementation of the correction factor, N_{Cor}, during a calculation. To do that, a correction factor is used as:

$$\eta = \frac{N_{Cor}}{K_p} \times \frac{P}{\rho g} \tag{6.23}$$

and

$$S_{\eta\eta} = \left(\frac{N_{Cor}}{K_p}\right)^2 \times \frac{S_{PP}}{\rho^2 g^2} \tag{6.24}$$

where, N_{Cor} is a correction factor which helps to account for a deviation of the measured wave profile from the linear wave theory (see Bishop and Donelan, 1987 for detail).

Despite the fact that a wave profile might not follow the linear wave theory in many cases, it has been shown by a series of studies that using a pressure response factor is adequately accurate if wave data are analyzed in the frequency domain. In other words, implementation of $N_{Cor} \approx 1$ leads to accurate outcomes (within 5%) in the frequency domain wave data analysis (e.g. Bishop and Donelan, 1987; Kuo and Chiu 1994; Townsend and Fenton, 1996; Tsai et al., 2005; Jones and Monismith, 2007). Nevertheless, an accurate evaluation of the final results from this method is required to guarantee validity of the outcomes.

6.8 Wave spectrum diagnostic tail

There might be situations that a high-frequency region of the measured power spectral density is missing or is not reliable. In other words, a tail section of the spectrum is missing or is not reliable. For example, such a condition can occur if water surface oscillation data are collected with a low sampling frequency. Another situation, which most likely causes a problem with a spectrum tail, can occur when data are collected by a pressure sensor deployed relatively deep under a water surface compared with a wavelength of the surface wave. It is explained that, when a pressure transducer is used to measure wave activities, shorter wave effects may not reach a depth that a pressure sensor is located. In such cases, the high-frequency wave energies may completely dissipate and diminish in the water column before reaching a depth that a pressure sensor is located. This makes a pressure sensor unable to detect the pressure effects of those high-frequency waves, i.e. short waves. In this case, if there is any measurement within that high-frequency range, it is most likely noise which has an insignificant signal-to-noise ratio. In any of these cases, all those high-frequency energies are lost and need to be replaced to avoid an under-estimation of the wave energy calculated from a water surface elevation power spectral density (e.g. Smith et al. 2001; Smith, 2002; Jones and Monismith, 2007).

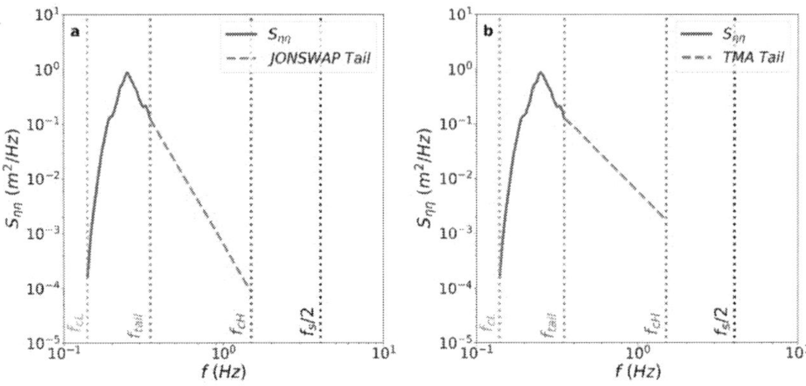

Figure 6.9 Replacing a power spectral density tail with a spectrum diagnostic tail. Plot (a) shows a schematic one-sided power spectral density which its tail is replaced by the JONSWAP diagnostic tail, with a tail power coefficient $n = 5$. Plot (b) shows a schematic one-sided power spectral density which its tail is replaced by the TMA diagnostic tail, with a tail power coefficient $n = 3$, and water depth $h = 5\ m$. In both plots, values with a frequency range of $f_{cL} \leq f \leq f_{cH}$ are retained and all values with a frequency range of $f < f_{cL}$ or $f > f_{cH}$ are removed. A low cut-off frequency is $f_{cL} = 0.14\ Hz$, a high cut-off frequency is $f_{cH} = 1.5\ Hz$, a sampling frequency is $f_s = 8\ Hz$, and $f_{tail} = 0.35\ Hz$. Values of f_{cL}, f_{cH}, f_s and f_{tail} in the figure are selected for a demonstration purpose. In practice, these values need to be selected based on the physical properties of the study area, instrumentation, and nature of the research. A schematic power spectral density is obtained from the manipulated JONSWAP spectrum (Hasselmann et al., 1973) for a wind velocity of $10\ m/s$, a wind fetch of $10000\ m$, and a peak wave frequency of $0.25\ Hz$, contaminated with randomly generated noise.

Regardless of the reason behind an unreliable spectrum tail, after a water surface elevation power spectral density is calculated either from surface elevation data or from dynamic pressure data, a high-frequency region of the water surface elevation spectrum should be treated to avoid error in the data analysis, particularly to prevent an under-estimation of the wave energy calculated from a water surface elevation spectrum. For example, after a wave spectrum is calculated from dynamic pressure data, to compensate for the wave energies that are not detected by a pressure sensor in the frequency domain, first, a simple high-cutoff frequency, f_{tail}, or a low-pass filter, is defined. The f_{tail} defines a frequency where spectrum data with $f > f_{tail}$ are unreliable. This high-cutoff frequency is used to remove a high-frequency part of the spectrum. In other words, f_{tail} is used to remove spectrum values for a frequency range of $f > f_{tail}$, which are most likely noises. After all noises are screened out, the lost high-frequency energies, which have a frequency range of $f > f_{tail}$ need to be replaced. To do that, an empirical spectrum tail, a so-called diagnostic tail, is used to replace values with a frequency range of $f > f_{tail}$ (e.g. Smith, 2002; Jones and Monismith, 2007; Siadatmousavi et al. 2012).

Depending on the measurement environment, a proper spectrum tail from an empirical wave spectrum is used to replace a tail of the measured water surface elevation spectrum. Commonly, one of the two well-known spectra, i.e. JONSWAP spectrum (Hasselmann et al., 1973) or TMA spectrum (Bouws *et al.*, 1985), is used for this purpose. For data collected in deep water, a noisy section of the spectrum tail is usually replaced with the JONSWAP spectrum tail (Hasselmann et al., 1973), while for data collected in intermediate and shallow water, a noisy section of the spectrum tail is commonly replaced with the TMA spectrum tail (Bouws *et al.*, 1985). A high-frequency section of the water surface elevation power spectral density is replaced by the JONSWAP diagnostic tail proportional to f^{-n} as (Siadatmousavi et al. 2012):

$$S_{\eta\eta}(f) = S_{\eta\eta}(f_{tail}) \times \left(\frac{f}{f_{tail}}\right)^{-n} \qquad \text{for } f > f_{tail} \tag{6.25}$$

Similarly, a high-frequency section of the water surface elevation power spectral density is replaced by the TMA spectrum tail proportional to f^{-n} as:

$$S_{\eta\eta}(f) = S_{\eta\eta}(f_{tail}) \times \frac{\Phi(f,h)}{\Phi(f_{tail},h)} \times \left(\frac{f}{f_{tail}}\right)^{-n} \qquad \text{for } f > f_{tail} \tag{6.26}$$

where, $S_{\eta\eta}(f_{tail})$ is a value of the measured spectrum, i.e. $S_{\eta\eta}$, at the frequency f_{tail}, the Φ is a transformation function, and n shows a tail power.

The f_{tail} is a frequency where all values of the measured spectrum with $f > f_{tail}$ are replaced by an empirical spectrum diagnostic tail. There are recommendations to set a value of f_{tail} equal to $f_{tail} = 2.5 f_m$ (Ardhuin et al. 2010) or equal to $f_{tail} = max(2.5 f_m, 4 f_{mPM})$ (Siadatmousavi et al. 2012), where $f_m = 1/T_{m01}$ is a mean wave frequency of the measured spectrum, f_{mPM} is a mean wave frequency of the Pierson-Moskowitz spectrum (Pierson and Moskowitz, 1964), $T_{m01} = m_0/m_1$ is a mean wave period, and $m_n = \int_0^\infty S_{\eta\eta}(f) f^n df$ is the nth moment of the wave energy spectrum, e.g. m_0 is a zero-moment of the wave energy spectrum. However, these recommendations might not be suitable for data that are collected in intermediate and shallow water and might result in a larger than adequate value for f_{tail} in depth-limited conditions. Therefore, for such depth-limited conditions, it is recommended to use a f_{tail} value in a range of $f_p < f_{tail} < 1.75 f_m$. For data that are collected by a pressure sensor, f_{tail} is recommended to be set equal to $f_{tail} = f_{maxpcorr}$, because data with a frequency range of $f_{maxpcorr} < f \leq f_{cH}$ might not be reliable. For any given condition, a visual evaluation of the power spectral density might be required to assure the chosen value for f_{tail} is acceptable.

The n in f^{-n} shows a tail power coefficient and defines a slope of the spectrum tail. A value of n depends on physical properties of the water body that data are collected in that. However, typically, a value of n is set equal to 5 for deep and 3 for shallow water (e.g. Phillips, 1958; Thornton, 1977; Kitaigorodskii, 1983; Miller and Vincent, 1990; Smith, 2002; Jones and Monismith, 2007; Holthuijsen, 2007; Kaihatu et al. 2007; Siadatmousavi et al. 2012). Therefore, a value of $n = 4$ might be suitable for intermediate depth.

The Φ is a transformation function. It transforms the JONSWAP spectrum in deep water into the TMA spectrum in depth-limited water as (Kitaigordskii et al., 1975; Hughes, 1984; Bouws et al., 1985; Holthuijsen, 2007):

$$\Phi = \frac{R_{\omega_h}^{-2}}{1 + 2\omega_h^2 R_{\omega_h} / \sinh\left(2\omega_h^2 R_{\omega_h}\right)} \tag{6.27}$$

where, R_{ω_h} and ω_h are:

$$R_{\omega_h} \tanh\left(\omega_h^2 R_{\omega_h}\right) = 1 \tag{6.28}$$

$$\omega_h = 2\pi f \sqrt{\frac{h}{g}} \tag{6.29}$$

A value of Φ can be calculated from (Tucker, 1994; Holthuijsen, 2007):

$$\Phi = \frac{\tanh^2(kh)}{1+2kh/\sinh(2kh)} \tag{6.30}$$

where, k is a wavenumber and h is water depth. Also, a value of Φ can be approximated by (Thompson and Vincent, 1983; Bergdahl, 2009):

$$\Phi(f,h) \approx \frac{\omega_h^2}{2} \qquad \text{for} \qquad \omega_h \leq 1 \tag{6.31}$$

$$\Phi(f,h) \approx 1 - 0.5(2-\omega_h)^2 \qquad \text{for} \qquad 1 < \omega_h < 2 \tag{6.32}$$

$$\Phi(f,h) = 1 \qquad \text{for} \qquad \omega_h \geq 2 \tag{6.33}$$

In general, it is not recommended to implement a diagnostic tail to replace a spectrum tail of the measured data. However, there might be situations that a tail of the measured spectrum is required to be replaced. For example, such situations that a high-frequency data are missing might occur if data are collected with a low sampling frequency, or data are collected with a pressure sensor that is deployed too deep in the water column. Also, if the high-frequency data are contaminated by noise, it needs to be replaced by a diagnostic tail.

7 Velocity Data Analysis

In this chapter, calculation of the wave properties from measured water particle orbital velocity data is covered. First, orbital velocity data conversion to a water surface elevation is explained. Next, estimation of the wave properties in the time and frequency domains is discussed. Then, a procedure for using a wave spectrum diagnostic tail is described. Step by step method to calculate wave properties from orbital velocity data in the frequency domain is presented.

Note that this chapter is similar to the pressure data analysis chapter. The orbital velocity data analysis has a similar procedure with the pressure data analysis, except for equations that need to be used. Therefore, a procedure from one can be transferred and used for another one by considering the differences. Although these two data analysis methods follow a similar procedure, for consistency in the provided data analysis techniques, the velocity data analysis is discussed separately and in details in this chapter.

7.1 Wave orbital and mean velocity (current velocity)

It very common to measure underwater velocity in coastal and oceanography studies by using instruments such as Acoustic Doppler Velocimeter (ADV) and Acoustic Doppler Profiler. Usually, collected velocity data consist of two main signals. The first signal shows a mean current velocity and the second signal shows a water particle motion orbital velocity. A mean current velocity represents a mass transport such as Stokes' drift and undertow. This mean current can be caused by physical phenomena such as nonlinear progressive waves (skewness and/or asymmetry of waves) and wave breaking. Orbital velocity is a water particle motion velocity caused by wave motion. In other words, water surface fluctuations or wave activities on a water surface cause water particle fluctuations under a water surface (e.g. Nielsen, 1992; Dronkers, 2005; Ciavola and Coco, 2017). Both mean and orbital velocities are used in coastal studies such as a coastal sediment transport. Furthermore, an orbital velocity can be used to calculate wave properties, which is a focus of this chapter.

For the purpose of data analysis, it is described that a time series is commonly split into a series of data blocks (data bursts) with a specific time interval. For example, data might split into a series of 1-hour bursts. Following the same procedure, it is required to split a velocity time series into a series of data bursts first. Then, the total measured velocity in each burst needs to be split into a mean velocity and orbital velocity. To separate these two velocities, first, a mean current velocity is calculated by averaging measured velocity data over time. Then, orbital velocity is obtained by subtracting a mean current velocity from the original measured velocity data. Therefore, a mean current velocity is equal to a mean value of the velocity data in each burst of the data, and orbital velocity is equal to the de-trended velocity data in that burst. Remember that data in each burst should be stationary for this purpose.

After orbital velocity data are obtained by de-trending the original velocity data, then they need to be prepared for the wave analysis. A reason that orbital velocity cannot be used right away after measurement for the wave analysis is that orbital velocity, which is resulted from water surface fluctuations, decrease in the water column as depth increases (e.g. Reeve et al., 2012). In other words, if a velocity probe is moving down from a water surface towards a seabed, it records smaller and smaller orbital velocity values for the similar wave as it moves down. If a velocity probe keeps moving down, eventually it reaches a depth where orbital velocity is almost damped out, and because of that, a sensor cannot sense orbital velocity anymore. In such conditions, a velocity probe only detects and records a mean velocity.

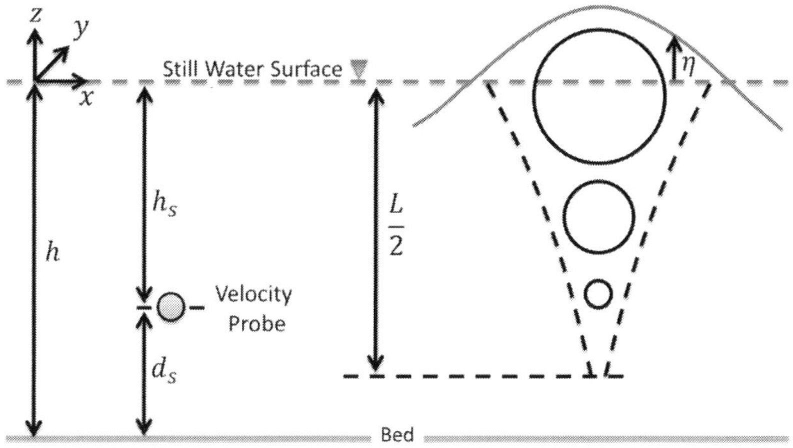

Figure 7.1 Schematic drawing of the water particles orbital motion and a velocity probe deployment in deep water.

The consequence of the orbital velocity attenuation in water depth is that orbital velocity recorded by a velocity probe is smaller and weaker compared with the values of the orbital velocity right beneath a water surface. Using these smaller orbital velocity data for wave analysis will lead to an underestimation of the wave height and energy. To get wave data from orbital velocity, first, the measured orbital velocities should be converted to water surface elevation. In a procedure of converting the orbital velocity data collected by a velocity probe to water surface elevation, the orbital velocity attenuation in water depth should be accounted for in order to get reliable results from data analysis.

To describe a procedure of the orbital velocity data conversion to water surface elevation, first, the measured horizontal velocity data are written as a combination of the mean current horizontal velocity and the horizontal orbital velocity as:

$$U_T(z = -h_s) = U_c + U \tag{7.1}$$

where, U_T is the total measured horizontal velocity at $z = -h_s$, the U_c is a mean part of the measured horizontal velocity at $z = -h_s$ which is equal to a mean current horizontal velocity, and U is a de-trended part of the measured horizontal velocity at $z = -h_s$ which is equal to the horizontal component of the water particle orbital velocity resulted from water surface fluctuations. The z is an upward vertical axis with zero at a water surface, h_s is a probe depth which is equal to $z = -h_s$. Commonly, a horizontal velocity and consequently horizontal orbital velocity is recorded in two perpendicular components, i.e. u in a x direction and v in a direction, where x and y are both in a horizontal plane. In such cases, U is a resultant horizontal orbital velocity and is equal to $U = (u^2 + v^2)^{0.5}$. Based on the linear wave theory, the horizontal orbital velocity of the water particle is:

$$U(f, z = -h_s) = \frac{\pi H}{T} \times \frac{\cosh k(h+z)}{\sinh(kh)} \times \cos(kx - \omega t) \tag{7.2}$$

or

$$U(f, z = -h_s) = \frac{\pi H}{T} \times \frac{\cosh(kd_s)}{\sinh(kh)} \times \cos(kx - \omega t) \tag{7.3}$$

where, H is wave height, h is local water depth, d_s is a velocity probe location distance from a seabed, $f = 1/T$ is a wave frequency, k is a wavenumber, T is a wave period, x is a distance, and t is time. A wavenumber is a function of the wave frequency, $f = 1/T$, and local water depth, h. A wavenumber is derived from a water wave dispersion relationship, i.e. $\omega^2 = (2\pi f)^2 = gk\tanh(kh)$, where $\omega = 2\pi f = 2\pi/T$ is an angular frequency.

Considering the linear wave profile as $\eta = \frac{H}{2} \times \cos(kx - \omega t)$, where η is a water surface elevation, then these relationships can be written as a function of η as:

$$U(f, z = -h_s) = \frac{2\pi}{T} \times \frac{\cosh k(h+z)}{\sinh(kh)} \times \eta \tag{7.4}$$

or

$$U(f, z = -h_s) = \frac{2\pi}{T} \times \frac{\cosh(kd_s)}{\sinh(kh)} \times \eta \tag{7.5}$$

Therefore, an orbital velocity conversion factor to water surface elevation is defined as:

$$K_{uv}(f, z = -h_s) = \frac{2\pi}{T} \times \frac{\cosh k(h+z)}{\sinh(kh)} = \frac{2\pi}{T} \times \frac{\cosh(kd_s)}{\sinh(kh)} \tag{7.6}$$

where, K_{uv} is a water particle horizontal orbital velocity to water surface elevation conversion factor.

After orbital velocity obtained from measured velocity data and an orbital velocity conversion factor is calculated, orbital velocity can be converted to water surface elevation. To achieve this, orbital velocity data are divided by an orbital velocity conversion factor. Therefore, a water surface elevation, η, is calculated as:

$$\eta = \frac{U}{K_{uv}} \quad (7.7)$$

where, η is a water surface elevation. Now, these water surface elevation data can be used for wave data analysis in the time or frequency domains.

In the frequency domain, a power spectral density of the water surface elevation can be obtained from the water surface calculated above. Alternatively, a power spectral density of the water surface elevation can be calculated from a power spectral density of the measured orbital velocity data as (Wiberg and Sherwood, 2008; Karimpour et al, 2016):

$$S_{\eta\eta} = \frac{S_{uv}}{K_{uv}^2} \quad (7.8)$$

where, $S_{\eta\eta}$ is a water surface elevation power spectral density, $S_{uv} = S_{uu} + S_{vv}$ is a combined horizontal orbital velocities power spectral density, and S_{uu} and S_{vv} are a power spectral density of the horizontal orbital velocity in x and y directions, respectively.

However, in practice, calculating a water surface elevation in the time domain and its spectrum in the frequency domain from the measured velocity is not a straightforward procedure and requires a series of steps before the final results obtained. In following sections, these procedures will be discussed in details.

7.2 Obtaining water surface elevation from measured velocity data in the time domain

One method to obtain a water surface elevation from measured velocity data is to analyze data in the time domain. To obtain a water surface elevation from measured orbital velocity, an orbital velocity conversion factor, K_{uv}, is required to be defined. The K_{uv} is a function of the water depth, h, and the wavenumber, k. Therefore, to calculate K_{uv}, water depth and a wavenumber are needed to be defined first. Water depth can be measured directly or obtained from other sensors such a pressure transducer. Some of the water velocity measurement instruments have an embedded pressure transducer which can be used to find water depth.

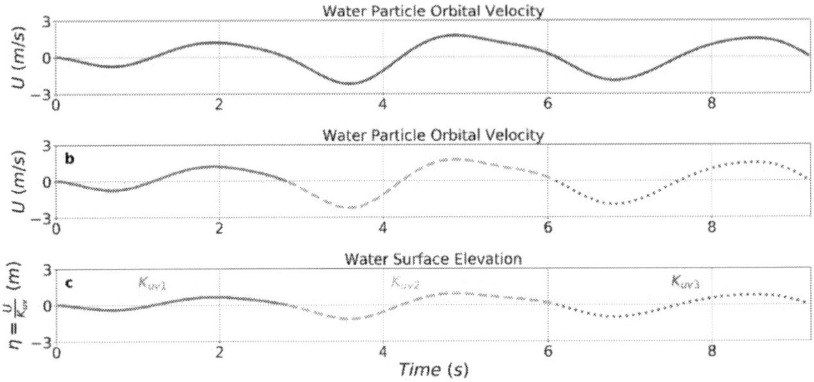

Figure 7.2 Converting water particle orbital velocity to water surface elevation. Plot (a) shows orbital velocity at a depth of 1 m below a water surface, i.e. $h_s = 1\ m$. Plot (b) shows orbital velocity is split into a series of individual waves by using a zero-crossing method. In a plot (c), every individual wave in a plot (b) is divided by an associated orbital velocity conversion factor, K_{uv}, to attain a water surface elevation, η. Here a zero down-crossing is used, however, a zero up-crossing can be used as well. Synthetic orbital velocity values in a plot (a) are obtained by generating a wave time series and multiplying it by an orbital velocity conversion factor. A synthetic wave time series is resulted from a superposition of the second-order Stokes' waves with amplitudes, periods, and phases equal to $a = 0.5, 0.9, 0.2\ m$, $T = 4, 3, 2\ s$, and $\phi = 0.55\pi, -0.55\ \pi, 0.55\pi\ rad$, respectively. Water depth is equal to $h = 5\ m$. Note that only the first three waves are shown.

To calculate a wavenumber, a wave period needs to be defined first. However, a calculation of the wave period is not as straightforward as a calculation of the water depth. Before calculating a wave period, first, it is considered that orbital velocity attenuation in the water column only has an effect on a velocity magnitude, and has not any effect on an orbital velocity oscillation period. In other words, a period of the measured orbital velocity oscillation under a water surface is the same as a wave period of the water surface oscillation at a water surface. By considering this fact, to obtain a wave period, first, measured orbital velocity data are split into a series of single waves by using a zero-crossing method. Then, a wave period for every single wave is defined. From that, a wavenumber and an orbital velocity conversion factor for each of those waves are calculated. After a K_{uv} is defined for each wave, the U values corresponding to each of the split waves are converted to water surface elevation by using the associated K_{uv} calculated for that wave as:

$$\eta = \frac{U}{K_{uv}} \tag{7.9}$$

After a water surface elevation corresponding to each of the split waves is calculated, an entire time series of the water level is obtained by reassembling and putting a water surface elevation associated with each split wave together. In other words, every single wave is converted to water surface elevation, and then all of these waves are put back together in the same order that they are in a measured velocity dataset. Finally, wave properties are obtained from this water surface elevation, η, by using a zero-crossing method.

Note that K_{uv} value approaches toward zero as a frequency of the wave increases, i.e. in short waves. As K_{uv} approaches toward zero, $1/K_{uv}$ approaches toward infinity. Therefore, to prevent an unacceptable magnification of the results as K_{uv} approaches toward zero, a minimum value for K_{uv} should be considered, otherwise it may result in very large values for very small K_{uv} (For detail, see velocity data analysis in the frequency domain).

7.3 Obtaining a power spectral density of the water surface elevation from measured velocity data in the frequency domain

It is more common to analyze wave data in the frequency domain rather than in the time domain. In other words, spectral analysis of wave data is a preferable method in most cases. One method to do that is to obtain a water surface elevation time series from velocity data in the time domain, and then use this surface elevation time series for the spectral analysis. However, it is not a usual practice as it is more common to directly calculate a wave spectrum from an orbital velocity spectrum. However, spectral analysis of the measured orbital velocity data can be very challenging.

In spectral analysis, similar to the time domain procedure, an orbital velocity conversion factor is used to convert orbital velocity to water surface elevation. However, this time, all calculations are carried out in the frequency domain. To do that, first, an orbital velocity conversion factor is calculated for all frequencies in a range of $0 \leq f \leq f_s/2$, where f_s is a sampling frequency. Then, a power spectral density of the water surface elevation, $S_{\eta\eta}$, is calculated by using a combined horizontal orbital velocities power spectral density, S_{uv}, as:

$$S_{\eta\eta}(f) = \frac{S_{uv}(f)}{\left(K_{uv}(f)\right)^2} \tag{7.10}$$

where, $S_{uv}(f) = S_{uu}(f) + S_{vv}(f)$. Then, wave properties are calculated from $S_{\eta\eta}$.

When data are analyzed in the frequency domain, it is necessary to remove all unreliable and noisy data from a spectrum. To remove these unreliable data, a low-cutoff frequency (a so-called high-pass filter), f_{cL}, or a high-cutoff frequency (a so-called low-pass filter), f_{cH}, or both f_{cL} and f_{cH} are used. In practice, f_{cL} and f_{cH} limits may be simply implemented by removing all data that have frequencies smaller than f_{cL}, i.e. $f < f_{cL}$ or have frequencies larger than f_{cH}, i.e. $f > f_{cH}$, from a dataset. In such cases, no transition band is considered between a passband and a stopband.

As noted above, when K_{uv} approaches toward zero, $1/K_{uv}$ approaches toward infinity. Therefore, similar to the time domain analysis, a minimum value for a K_{uv} should be considered to prevent an unacceptable magnification of the results as K_{uv} approaches toward zero, otherwise, it may result in very large values for very small K_{uv}.

7.4 Lower limit for an orbital velocity conversion factor

As mentioned above, in theory, a water surface elevation, η, time series in the time domain or a water surface elevation power spectral density, $S_{\eta\eta}$, in the frequency domain can be obtained by using an orbital velocity conversion factor, K_{uv}. However, in practice, an important question needs to be addressed first in order to get reliable values from these methods in the time and frequency domains. The question is: What is the smallest K_{uv} value that can be used for a given dataset?

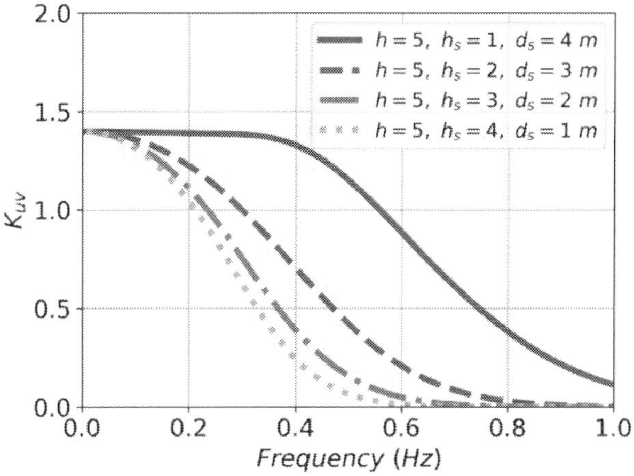

Figure 7.3 Schematic trend of the orbital velocity conversion factor, K_{uv}, versus a frequency. The K_{uv} is calculated for water depth equal to $h = 5\ m$ and four sensor depths equal to $h_s = 1, 2, 3, 4\ m$. Associated sensor heights from a seabed are $h_s = 4, 3, 2, 1\ m$, respectively. Note that only a frequency range of $0 \leq f \leq 1\ Hz$ is plotted.

A reason that an answer to this question is crucial is that, in theory, an orbital velocity conversion factor, K_{uv}, which is a function of the wave frequency (wave period), starts to decrease as a frequency increases in a higher frequency region of the spectrum. This results in very small values for K_{uv} in higher frequencies. Eventually, as a frequency increases further, K_{uv} values approach zero. These small values for K_{uv} (large values for $1/K_{uv}$) lead to an unrealistic magnification of the $S_{\eta\eta}$ values in higher frequencies. In other words, the outcomes of $\eta = U/K_{uv}$ in the time domain, and $S_{\eta\eta} = S_{uv}/(K_{uv})^2$ in the frequency domain become unrealistically large when K_{uv} approaches to zero for short waves. This unrealistic amplification of the short waves can lead to an error and unacceptable magnification in a calculation of the water level fluctuation or wave energy.

To prevent an error caused by small K_{uv} values, a high-cutoff frequency, $f_{maxuvcorr}$, which is associated with a minimum allowable orbital velocity conversion factor, K_{uvmin}, is chosen to limit a minimum value of the K_{uv} to K_{uvmin}. It means that for frequency values larger than this cutoff frequency, i.e. for $f \geq f_{maxuvcorr}$, values of K_u become smaller than K_{uvmin}, and therefore,

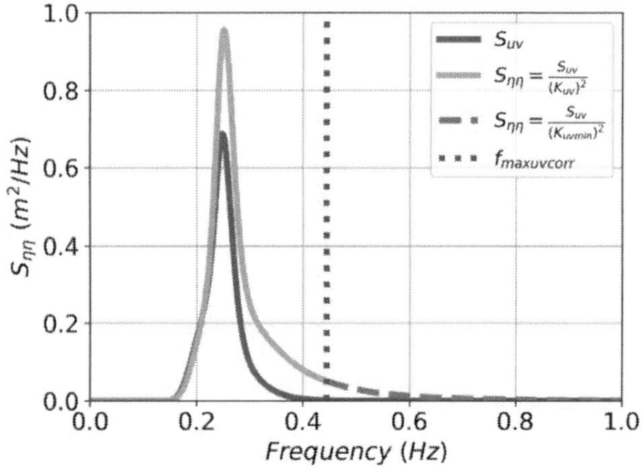

Figure 7.4 Schematic illustration of obtaining a water surface elevation power spectral density, $S_{\eta\eta}$, from a combined horizontal orbital velocities power spectral density, S_{uv}. Plot shows $f_{maxuvcorr}$ where a value of K_{uv} is limited to K_{uvmin} for a frequency range of $f \geq f_{maxuvcorr}$. A schematic power spectral density is obtained from the JONSWAP spectrum (Hasselmann et al., 1973) for a wind velocity of 10 m/s, a wind fetch of 10000 m, and a peak wave frequency of 0.25 Hz.

they should be limited to K_{uvmin}. In other words, if $K_{uv} < K_{uvmin}$ then $K_{uv} = K_{uvmin}$. Limiting a minimum of the K_{uv} value to K_{uvmin} prevents amplification of the small waves or even noise with high frequencies. Therefore, this helps to avoid an error due to an over-estimation of the water surface oscillation or wave energy.

7.5 Theoretical method to define an acceptable minimum value for an orbital velocity conversion factor

It is shown that for any given dataset, a minimum value for an orbital velocity conversion factor, K_{uvmin}, and its associated frequency, $f_{maxuvcorr}$, needs to be defined. To define K_{uvmin}, first, the smallest wave that its effect reaches the velocity probe deployment depth needs to be defined. To find this smallest wave, the linear wave theory can be used.

The linear wave theory indicates that for any given wave, water particle motions resulted from water surface wave activities dissipate as water depth increase, and they nearly diminish in water depth larger than half of the wavelength. In other words, water surface fluctuation effects (wave effects) are only detectable from a water surface down to a depth equal to half of the wavelength of the surface wave, i.e. where $h \leq L/2$, where L is a wavelength.

Therefore, following the linear wave theory, a velocity probe located at a depth equal to h_s can only detect waves with wavelengths that follow $L/2 \geq h_s$, and cannot sense waves with wavelengths that follow $L/2 \leq h_s$. It means that the smallest wave that is detectable by a velocity probe located at depth h_s, has a wavelength that satisfies $L/2 = h_s$. Because of that, all data with a wavelength that follows $L/2 \leq h_s$ are likely noise and should be removed from a dataset. Therefore, it is essential to define $f_{maxuvcorr}$ or K_{uvmin} accurately to ensure K_{uv} is only being applied to waves that are sensed by a velocity probe. This helps to prevent any amplification of the noise or waves that their effects do not reach a sensor's location.

Theoretically, values of K_{uv} and $f_{maxuvcorr}$ can be estimated by using the linear wave theory. As mentioned above, based on the linear wave theory, the water particle motion resulted from the wave oscillation is almost diminished at a depth equal to half of the wavelength of the surface wave. Therefore, a wavelength of the shortest linear wave, $L_{min-Linear}$, that is detectable by a velocity probe located at depth h_s follows:

$$h_s = \frac{L_{min-Linear}}{2} \tag{7.11}$$

where, L is a wavelength, $L_{min-Linear}$ is a wavelength of the shortest linear wave that is detectable by a velocity probe, h_s is a velocity probe location depth, and a subscript *Linear* denotes that the linear wave theory is used to estimate a parameter. Considering a wavenumber is $k = 2\pi/L$, this relationship can be written as:

$$h_s = \frac{\pi}{k_{max-Linear}} \tag{7.12}$$

or

$$k_{max-Linear} h_s = \pi \tag{7.13}$$

where, $k_{max-Linear}$ is a wavenumber of the shortest linear wave that is detectable by a velocity probe and is associated with $L_{min-Linear}$. Considering $h_s = h - d_s$, where d_s is a velocity probe distance from a seabed, this relationship can be re-written as:

$$k_{max-Linear}(h - d_s) = \pi \tag{7.14}$$

or

$$k_{max-Linear} = \frac{\pi}{h-d_s} \tag{7.15}$$

Then, a minimum acceptable value for an orbital velocity conversion factor, estimated from the linear wave theory, $K_{uvmin-Linear}$, is equal to:

$$K_{uvmin-Linear}(f_{maxuvcorr-Linear}, z = -h_s) = \frac{2\pi}{T} \times \frac{\cosh(k_{max-Linear} d_s)}{\sinh(k_{max-Linear} h)} \tag{7.16}$$

or

$$K_{uvmin-Linear}(f_{maxuvcorr-Linear}, z = -h_s) = \frac{2\pi}{T} \times \cosh\left(\frac{\pi}{h-d_s} d_s\right) / \sinh\left(\frac{\pi}{h-d_s} h\right) \tag{7.17}$$

or

$$K_{uvmin-Linear}(f_{maxuvcorr-Linear}, z = -h_s) = 2\pi f_{maxuvcorr-Linear} \times \cosh\left(\frac{\pi}{h-d_s} d_s\right) / \sinh\left(\frac{\pi}{h-d_s} h\right) \tag{7.18}$$

To estimate an associated frequency with $K_{uvmin-Linear}$, a water wave dispersion relationship is written as:

$$\omega^2 = (2\pi f)^2 = gk \tanh(kh) \tag{7.19}$$

where, $\omega = 2\pi f$ is a wave angular frequency, and k is a wavenumber. A dispersion relationship can be re-written as:

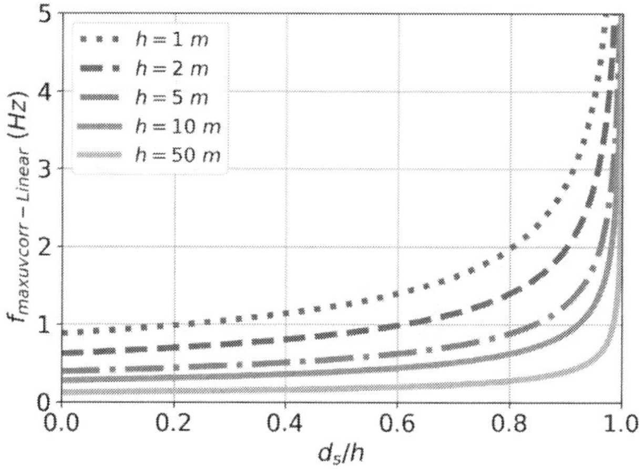

Figure 7.5 Trend of $f_{maxuvcorr-Linear}$ versus d_s/h for water depth equal to $h = 1, 2, 5, 10, 50\ m$. The $f_{maxuvcorr-Linear}$ shows a maximum frequency calculated from the linear wave theory beyond which $K_{uv} = K_{uvmin-Linear}$. Note that $f_{maxuvcorr-Linear} = f_{maxpcorr-Linear}$.

$$f = \frac{1}{2\pi}\sqrt{gk\tanh(kh)} \tag{7.20}$$

By replacing k with $k_{max-Linear}$, a frequency associated with $K_{uvmin-Linear}$ is:

$$f_{maxuvcorr-Linear} = \frac{1}{2\pi}\sqrt{gk_{max-Linear}\tanh(k_{max-Linear}h)} \tag{7.21}$$

or

$$f_{maxuvcorr-Linear} = \frac{1}{2\pi}\sqrt{g\frac{\pi}{h-d_s}\tanh\left(\frac{\pi}{h-d_s}h\right)} \tag{7.22}$$

where, $f_{maxuvcorr-Linear}$ is a frequency associated with $K_{uvmin-Linear}$. The $f_{maxuvcorr-Linear}$ shows a maximum frequency estimated by the linear wave theory where at that frequency $K_{uv} = K_{uvmin-Linear}$. Therefore, for frequencies larger than $f_{maxuvcorr-Linear}$, i.e. for $f \geq f_{maxuvcorr-Linear}$, values of K_{uv} become smaller than $K_{uvmin-Linear}$, and because of that, values of K_{uv} should be limited to $K_{uvmin-Linear}$ in this frequency range. It should be pointed out that a value of $f_{maxuvcorr-Linear}$ is equal to a value of $f_{maxpcorr-Linear}$.

If a velocity probe sits on a seabed, then $d_s = 0$ and $z = -h = -h_s$. Therefore:

$$k_{max-Linear} h_s = k_{max-Linear} h = \pi \tag{7.23}$$

Then, $K_{uvmin-Linear}$ for a velocity probe that sits on a seabed is:

$$K_{uvmin-Linear}(f_{maxuvcorr-Linear}, z = -h) = \frac{2\pi}{T} \times \frac{\cosh(0)}{\sinh(\pi)} \approx 0.544T^{-1} \tag{7.24}$$

or

$$K_{uvmin-Linear}(f_{maxuvcorr-Linear}, z = -h) \approx 0.544 f_{maxuvcorr-Linear} \tag{7.25}$$

And, $f_{maxuvcorr-Linear}$ for a velocity probe that sits on a seabed is:

$$f_{maxuvcorr-Linear} = \frac{1}{2\pi}\sqrt{g\frac{\pi}{h}\tanh(\pi)} \approx 0.8819h^{-0.5} \tag{7.26}$$

7.6 Practical methods to define an acceptable minimum value for an orbital velocity conversion factor

It is explained how to use the linear wave theory to predict $f_{maxuvcorr-Linear}$ and $K_{uvmin-Linear}$ values. However, these theoretical values are only applicable to sinusoidal waves. It means that if measured orbital velocity data are following a sinusoidal profile, then considering $f_{maxuvcorr} \approx f_{maxuvcorr-Linear}$ and $K_{uvmin} \approx K_{uvmin-Linear}$ leads to satisfactory results.

However, if a wave profile departs from a sinusoidal form then using $f_{maxuvcorr-Linear}$ and $K_{uvmin-Linear}$ in the data analysis is not acceptable anymore. Such conditions may occur when waves propagate into intermediate and shallow water depth. Under these conditions, a waveform becomes skewed and does not follow a sinusoidal form anymore. When a waveform deviates from a sinusoidal form, then $f_{maxuvcorr} <$

$f_{maxuvcorr-Linear}$ and $K_{uvmin} > K_{uvmin-Linear}$. It means that considering $f_{maxuvcorr} \approx f_{maxuvcorr-Linear}$ and $K_{uvmin} \approx K_{uvmin-Linear}$ for non-sinusoidal waves will result in an over-estimation of the water surface oscillation and wave energy.

In practice, $f_{maxuvcorr}$ and K_{uvmin} need to be defined as accurately as possible to avoid signal inflation where $f > f_{maxuvcorr}$. To do that, different techniques can be implemented in the time and frequency domains.

7.6.1 Practical method to define $f_{maxuvcorr}$ and K_{uvmin} in the time domain

To use K_{uvmin} in the time domain data analysis, first, a constant predefined K_{uvmin} is chosen. This would be the lowest value K_{uv} can get in the analysis. For example, a constant predefined $K_{uvmin} = 0.15$ is a justifiable choice. Considering a $K_{uvmin} = 0.15$ helps to prevent unacceptable amplification of the measured data greater than 6.66 times (i.e. $1/0.15$ times). Note that, although $K_{uvmin} = 0.15$ might be suitable for most cases, this value need to be chosen based on the waves and water body conditions.

After this K_{uvmin} value is set, all waves in a time series are split into a series of individual waves. Then, $K_{uvmin-Linear}$ is calculated for each of these individual waves. Finally, a calculated $K_{uvmin-Linear}$ for each wave is compared with a constant predefined K_{uvmin}, and a maximum value between them, i.e. $max(K_{uvmin-Linear}, K_{uvmin})$, is used to convert velocity data to water surface elevation for that wave. This procedure is repeated for all individual waves in a time series.

7.6.2 Practical methods to define $f_{maxuvcorr}$ and K_{uvmin} in the frequency domain

In the frequency domain data analysis, there are three approaches that can be implemented to define $f_{maxuvcorr}$ and K_{uvmin}, which helps to avoid signal inflation where $f > f_{maxuvcorr}$.

In the first approach, first, a combined horizontal orbital velocities power spectral density, S_{uv}, is calculated. Then, a noise floor at the end of the power spectral density is defined. Next, this noise floor is uniformly subtracted from a power spectral density. Finally, a water surface elevation power spectral density is calculated from $S_{\eta\eta} = S_{uv}/K_{uv}^2$ with its noise floor removed. In this approach, it is assumed that a noise floor is a uniformly distributed wide band noise for all frequencies. This approach reflects an assumption that a frequency that a noise floor starts at the tail of the power spectral density is equal to $f_{maxuvcorr}$. In other words, a frequency that a noise floor emerges at a tail of the power spectral density is a frequency of the shortest wave that a velocity probe can detect. Therefore, the effects of the shorter waves with larger wave frequencies cannot be detected by a velocity probe as they diminished before reaching a sensor depth. Because of that, all the readings appear after $f_{maxuvcorr}$ are considered as noise.

In the second approach, which is more common than the first one, first, a high-cutoff frequency equal to f_{cH} is selected and assumed it is equal to $f_{maxuvcorr}$, i.e. it is assumed that $f_{maxuvcorr} = f_{cH}$. Then, all data with frequencies $f > f_{cH}$ are filtered out of the power spectral density. The f_{cH} is chosen in a way to assure removal of all data with a low signal-to-noise ratio. A simple method to define f_{cH} is to set f_{cH} equal to a frequency that a noise floor starts at a tail of the power spectral density. In this method, a high-cutoff frequency can be constant (the same) for all bursts in a dataset, or it can be selected adaptively for each burst. Between these two, a selection of the constant cutoff frequency, i.e. $f_{maxuvcorr} = f_{cH} =$ Constant, for all bursts is recommended. Note that, selecting an adaptive cutoff frequency, i.e. $f_{maxuvcorr} = f_{cH} =$ Adaptive, for each burst is not suggested as it can result in dissimilar conditions from burst to burst.

The third approach is a modified form of the second approach, aimed to improve the second approach for defining $f_{maxuvcorr}$ and K_{uvmin}. In this approach, two different values are chosen for f_{cH} and $f_{maxuvcorr}$, where $f_{maxuvcorr} \leq f_{cH}$. It is an acceptable practice in the frequency domain data analysis to use a constant high-cutoff frequency, i.e. $f_{cH} =$ Constant, in most cases. In contrast, values of $f_{maxuvcorr}$ and K_{uvmin} can be either constant for all bursts or can be selected adaptively for each burst. However, in most situations, it is inadequate to consider a constant value for $f_{maxuvcorr}$ and K_{uvmin}, because it can result in an incorrect wave estimation due to either an over or under-estimation of the wave energy. Therefore, in the third approach, a constant value is considered for f_{cH}, while adaptive values are implemented for $f_{maxuvcorr}$ and K_{uvmin}, unless using constant values for $f_{maxuvcorr}$ and K_{uvmin} are justifiable. In other words, while f_{cH} = Constant for the entire dataset, values of $f_{maxuvcorr}$ and K_{uvmin} either are $f_{maxuvcorr}$ = Adaptive and K_{uvmin} = Adaptive, which are defined for each burst individually, or $f_{maxuvcorr}$ = Constant and K_{uvmin} = Constant for the entire dataset. A constant value for f_{cH} guarantees a similar high-cutoff frequency for all wave spectrums in case that a series of wave spectrums are calculated for a series of bursts. Calculating the adaptive values for $f_{maxuvcorr}$ and K_{uvmin} in each burst help to avoid noise inflation, as a noise floor start-frequency may differ from burst to burst. Then, unlike the second approach that assumes $f_{maxuvcorr} = f_{cH}$, in the third approach $f_{maxuvcorr} \leq f_{cH}$.

7.6.3 Some guidelines on defining a constant f_{cH}, an adaptive $f_{maxuvcorr}$, and an adaptive K_{uvmin} in the frequency domain

If the third approach is selected for data analysis, it is recommended to use a constant value for f_{cH}, and adaptive values for $f_{maxuvcorr}$ and K_{uvmin}. To use this method, these values, i.e. a constant value for f_{cH}, and adaptive values for $f_{maxuvcorr}$ and K_{uvmin}, need to be defined. In this approach, if only one spectrum needs to be evaluated, then defining f_{cH} = Constant, $f_{maxuvcorr}$ = Adaptive and K_{uvmin} = Adaptive might not be very complicated. However, in most studies, a series of data bursts are collected, and a power spectral density needs to be calculated for each individual burst. In such conditions, which a large number of spectrums need to be calculated, defining f_{cH} = Constant that works for all spectrums and estimating $f_{maxuvcorr}$ = Adaptive and K_{uvmin} = Adaptive for each individual spectrum can be very challenging.

Selecting a constant f_{cH} value that suits all spectrums involves in finding a value for f_{cH} which eliminates noise at a tail of all power spectral densities. This commonly can be done by visually inspecting a series of S_{uv} spectrums before applying K_{uv} to convert them to water surface elevation. This helps to identify a single value for f_{cH} which filters out noise in all spectrums.

Defining adaptive values of $f_{maxuvcorr}$ and K_{uvmin} for each burst is not as straightforward. In general, a value of $f_{maxuvcorr}$ for each burst of data is defined by searching for an exact location of $f_{maxuvcorr}$ in a power spectral density of that burst. There are two methods to search and find the values of $f_{maxuvcorr}$ and K_{uvmin} in each burst.

In the first method, first, a combined horizontal orbital velocities power spectral density, S_{uv}, is calculated. Next, a noise floor at the end of the power spectral density is identified. Then, $f_{maxuvcorr}$ is set equal to a frequency that a noise floor emerges at a tail of the power spectral density. It is based on the assumption that $f_{maxuvcorr}$ represents a frequency of the shortest wave that a velocity probe can detect, and typically it is equal to a frequency where a noise floor starts at a tail of the power spectral density.

In the second method, first, a water surface elevation power spectral density is calculated by applying K_{uv} without limiting K_{uv} to K_{uvmin} as $S_{\eta\eta} = S_{uv}/K_{uv}^2$, where K_{uv} can approach toward zero without a limitation. By not limiting K_{uv} to K_{uvmin}, a calculated spectrum at first increases to its peak value and then it starts to decrease until it reaches to a minimum value. After dropping to this minimum point, it starts to rise again and moves toward infinity. A frequency of this minimum value in a spectrum tail before it moves toward infinity is $f_{maxuvcorr}$. Note that a location of this minimum is after a spectrum peak frequency, f_p, in a tail of the spectrum.

After a value of $f_{maxuvcorr}$ is defined by either of these methods, a minimum value of the orbital velocity conversion factor, K_{uvmin}, which is a K_{uv} value associated with $f_{maxuvcorr}$, is calculated. Then, an orbital velocity conversion factor is calculated for all frequencies in a range of $0 \leq f \leq f_s/2$. After that, the lower boundary of K_{uv} is limited to K_{uvmin} as $K_{uv} \geq K_{uvmin}$ for all frequencies, where $K_{uv} = K_{uvmin}$ at $f = f_{maxuvcorr}$ Hz.

Note that in this approach, the adaptive $f_{maxuvcorr}$ and K_{uvmin} should always be limited to values calculated from the linear wave theory as $f_{maxuvcorr} \leq f_{maxuvcorr-Linear}$ and $K_{uvmin} \geq K_{uvmin-Linear}$.

If higher harmonics of the dominant frequency have considerable energies, then a water surface elevation power spectral density might have multiple energy peaks which are associated with those higher harmonics. In particular, this may be a case when the 2nd harmonic is presented in the data. In such cases, additional attention requires to avoid identifying these higher harmonics mistakenly as a high-frequency amplification caused by the small

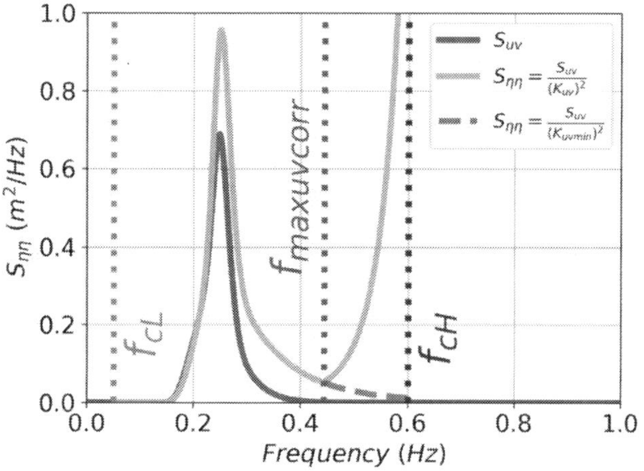

Figure 7.6 Schematic illustration of obtaining a water surface elevation power spectral density, $S_{\eta\eta}$, from a combined horizontal orbital velocities power spectral density, S_{uv}. Plot shows how high-frequency energy values are amplified if a value of K_{uv} is not limited by K_{uvmin}. To prevent artificial amplification of the spectrum, $f_{maxuvcorr}$ is chosen to limits K_{uv} to K_{uvmin} for a frequency range of $f \geq f_{maxuvcorr}$. A low cut-off frequency is $f_{cL} = 0.05\ Hz$, a high cut-off frequency is $f_{cH} = 0.6\ Hz$, and a sampling frequency is $f_s = 2\ Hz$. Values of f_{cL}, f_{cH} and f_s in the figure are selected for a demonstration purpose. In practice, these values need to be selected based on the physical properties of the study area, instrumentation, and nature of the research. A schematic power spectral density is obtained from the JONSWAP spectrum (Hasselmann et al., 1973) for a wind velocity of 10 m/s, a wind fetch of 10000 m, and a peak wave frequency of 0.25 Hz.

K_{uv} values. To do that, an adaptive $f_{maxuvcorr}$ needs to be chosen in a way that it reflects the higher harmonic effects without amplifying the noise.

After $f_{maxuvcorr}$ is defined, then it is compared with f_{cH}. If a value of $f_{maxuvcorr}$ becomes equal or larger than a constant value chosen for f_{cH}, i.e. if $f_{maxuvcorr} \geq f_{cH}$, then a value of $f_{maxuvcorr}$ is limited to f_{cH} as $f_{maxuvcorr} = f_{cH}$. Otherwise, if a value of $f_{maxuvcorr}$ is smaller than f_{cH}, i.e. if $f_{maxuvcorr} < f_{cH}$, then additional attention is required for applying K_{uv} within a frequency range of $f_{maxuvcorr} \leq f \leq f_{cH}$.

If $f_{maxuvcorr} < f_{cH}$, then two approaches may be considered to apply K_{uv} to the data within a frequency range of $f_{maxuvcorr} \leq f \leq f_{cH}$. In the first approach, a transition band is considered from $f_{maxuvcorr} - \Delta f_{tr}$ to $f_{maxuvcorr} + \Delta f_{tr}$, where $\pm\Delta f_{tr}$ defines a frequency range for a transition band before and after $f_{maxuvcorr}$ and is chosen based on the spectrum's conditions. Then, K_{uv} is gradually increased from $K_{uv} = K_{uvmin}$ at $f = f_{maxuvcorr} - \Delta f_{tr}$ to $K_{uv} = 1$ at $f = f_{maxuvcorr} + \Delta f_{tr}$ and retained equal to 1 for higher frequency, i.e. $K_{uv} = 1$ for $f_{maxuvcorr} + \Delta f_{tr} \leq f \leq f_{cH}$.

In the second approach, a value of K_{uv} is kept equal to K_{uvmin} for frequencies larger than $f_{maxuvcorr}$, i.e. $K_{uv} = K_{uvmin}$ for $f_{maxuvcorr} \leq f \leq f_{cH}$. A difference between these two approaches is that, in the second approach, where $K_{uv} = K_{uvmin}$ for $f_{maxuvcorr} \leq f \leq f_{cH}$, a slope of the spectrum tail obtained from $S_{\eta\eta} = S_{uv}/K_{uv}^2$ would be similar to a slope of spectrum tail obtained from S_{uv}. However, in the first approach, an increase of K_{uv} from $K_{uv} = K_{uvmin}$ to $K_{uv} = 1$ leads to dissimilar spectrum tail slopes for $S_{\eta\eta} = S_{uv}/K_{uv}^2$ compared to S_{uv}.

Finally, , after K_{uvmin} is defined and K_{uv} is calculated for all frequencies, a water surface elevation power spectral density, $S_{\eta\eta}$, is calculated from $S_{\eta\eta} = S_{uv}/K_{uv}^2$ where $K_{uv} \geq K_{uvmin}$.

7.6.4 Step by step procedure to obtain a water surface elevation power spectral density from measured velocity data in the frequency domain

All aforementioned steps are required to be followed to attain accurate results from the spectral analysis. However, following these steps can be very overwhelming, particularly for defining adaptive values of $f_{maxuvcorr}$ and K_{uvmin}. To provide a better guideline, a step by step procedure, which is applicable in most cases, including how to define and implement adaptive values of $f_{maxuvcorr}$ and K_{uvmin}, is described. To convert measured velocity data to water surface elevation in the frequency domain, it is recommended to follow these steps:

- Calculate a combined horizontal orbital velocities power spectral density as $S_{uv} = S_{uu} + S_{vv}$.
- Visually evaluate a series of S_{uv} spectrums (as many as needed) and define a minimum frequency where spectrum values with smaller frequencies are not reliable. Set this frequency equal to f_{cL}. This indicates that spectrum values with a frequency range of $f < f_{cL}$ are

not reliable. As this value of f_{cL} is used for all spectra, it might be required to evaluate a series of spectra to find a f_{cL} value that is applicable for all spectra that are intended to be analyzed.

- Visually evaluate a series of S_{uv} spectrums (as many as needed) and define a maximum frequency which is required for the data analysis. In other words, decide about the highest frequency limit where spectrum values should be retained up to that point in order to carry out this particular research. Set this frequency equal to f_{cH}. There might be data with a frequency range of $f_{maxuvcorr} \leq f \leq f_{cH}$ or $f_{tail} \leq f \leq f_{cH}$ where spectrum data are not reliable within that range. In such cases, either a value of f_{cH} should be decreased to a frequency that prevents these unreliable values retained in a spectrum, or these data need to be replaced by an empirical spectrum diagnostic tail. As this value of f_{cH} is used for all spectra, it might be required to evaluate a series of spectra to find a f_{cH} value that is applicable for all spectra that are intended to be analyzed.
- Visually evaluate a couple of S_{uv} spectrums (as many as needed) and define a frequency that a noise floor starts at a high-frequency range of the spectrum. A frequency that a noise floor starts is equal to $f_{NoiseFloor}$. Spectrum values with a frequency range of $f > f_{NoiseFloor}$ are not reliable. As this value of $f_{NoiseFloor}$ is used for all spectra, it might be required to evaluate a series of spectra to find a $f_{NoiseFloor}$ value that is applicable for all spectra that are intended to be analyzed.
- Define a peak wave frequency, f_p, of the spectrum. It is a frequency which is associated with a maximum value of the spectrum.
- Calculate an orbital velocity conversion factor, K_{uv}, for a frequency range of $0 \leq f \leq f_s/2$. A value of K_{uv} starts to decrease toward small values as a frequency increases in a high-frequency region of the spectrum. Usually, a minimum value of K_{uv} is obtained for $f = f_s/2$. If a sampling frequency, f_s, becomes large enough, then K_{uv} approaches zero.
- Calculate a water surface elevation power spectral density by applying K_{uv} without limiting K_{uv} to K_{uvmin} as $S_{\eta\eta} = S_{uv}/K_{uv}^2$ for a frequency range of $0 \leq f \leq f_s/2$. The K_{uv} value can have a range of $K_p \geq 0$ depending on a range of $0 \leq f \leq f_s/2$.

- Remove all values of the spectrum that have a frequency range of $f < f_{cL}$ and $f > f_{cH}$. In other words, only retain spectrum values with a frequency range of $f_{cL} \leq f \leq f_{cH}$.
- Find a minimum value of the spectrum in a frequency range of $f_p \leq f \leq f_{cH}$. Set this frequency equal to $f_{MinPoint}$.
- Set $f_{maxuvcorr}$ equal to a smaller value between a frequency associated with a minimum point after peak frequency, i.e. $f_{MinPoint}$, and a frequency that a noise floor starts, i.e. $f_{NoiseFloor}$. In other words, set $f_{maxuvcorr} = min(f_{MinPoint}, f_{NoiseFloor})$.
- From the linear wave theory, calculate $f_{maxuvcorr-Linear}$ and compare it with $f_{maxuvcorr}$. If $f_{maxuvcorr}$ is larger than $f_{maxuvcorr-Linear}$, then set $f_{maxuvcorr}$ equal to $f_{maxuvcorr-Linear}$. In other words, if $f_{maxuvcorr} > f_{maxuvcorr-Linear}$ then $f_{maxuvcorr} = f_{maxuvcorr-Linear}$. Remember that $f_{maxuvcorr} \leq f_{maxuvcorr-Linear}$ should be satisfied for all cases.
- Calculate an orbital velocity conversion factor, K_{uv}, associated with $f_{maxuvcorr}$ and set it equal to K_{uvmin}.
- Calculate a water surface elevation power spectral density by applying K_{uv}, but this time limit a K_{uv} value to K_{uvmin} as $S_{\eta\eta} = S_{uv}/K_{uv}^2$, where $K_{uv} \geq K_{uvmin}$ for a frequency range of $0 \leq f \leq f_s/2$.
- Remove all values of the spectrum that have a frequency range of $f < f_{cL}$ and $f > f_{cH}$. In other words, only retain spectrum values with a frequency range of $f_{cL} \leq f \leq f_{cH}$.
- If required, replace values of the spectrum tail with a frequency range of $f_{maxuvcorr} \leq f \leq f_{cH}$ or $f_{tail} \leq f \leq f_{cH}$ with an empirical wave spectrum diagnostic tail (see wave spectrum diagnostic tail for detail).

Note that if constant values for $f_{maxuvcorr}$ and K_{uvmin} are chosen, then the same procedure with excluding the steps that define adaptive values for $f_{maxuvcorr}$ and K_{uvmin} can be followed.

Although this step by step guideline is applicable in most cases, note that there might be cases that this guideline does not produce reliable results. Therefore, in any data analysis, outcomes should be evaluated accurately to assure the quality of final results.

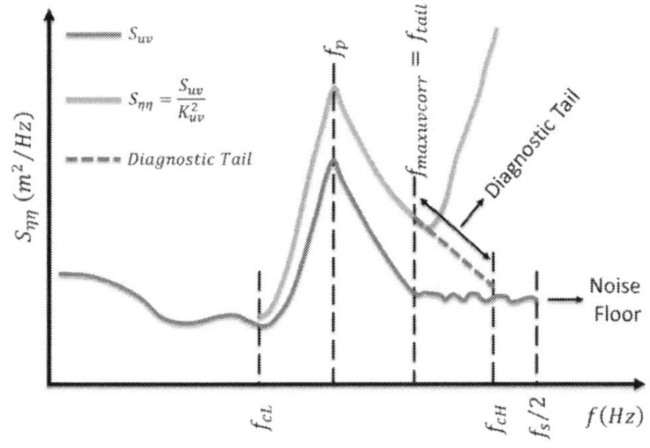

Figure 7.7 Schematic illustration of defining a low-cutoff frequency, f_{cL}, a high-cutoff frequency, f_{cH}, and $f_{maxuvcorr}$. Plot shows a schematic combined horizontal orbital velocities power spectral density, S_{uv}, within a frequency range of $0 \leq f \leq f_s/2$, and a water surface elevation power spectral density as $S_{\eta\eta} = S_{uv}/K_{uv}^2$ within a frequency range of $f_{cL} \leq f \leq f_{cH}$. In this plot, all values with a frequency range of $f_{cL} \leq f \leq f_{cH}$ are retained and all values with a frequency range of $f < f_{cL}$ or $f > f_{cH}$ are removed. The $f_{maxuvcorr}$ is selected equal to a frequency that a noise floor starts at a tail of the spectrum. The f_{tail} is chosen equal to $f_{maxuvcorr}$. A spectrum diagnostic tail is used for a frequency range of $f_{maxuvcorr} \leq f \leq f_{cH}$. All frequency values and a diagnostic tail in the figure are for a demonstration purpose. In practice, these values need to be selected based on the physical properties of the study area, instrumentation, and nature of the research. Similarly, a diagnostic tail might or might not be required.

7.7 Accuracy of using an orbital velocity conversion factor to convert velocity data to water surface elevation

A described method above, which use an orbital velocity conversion factor to convert velocity data to water surface elevation, is based on the linear wave theory. However, there are many instances in the real world that a wave profile does not follow the linear wave theory. A deviation of wave profile from the linear wave theory may raise concern about how accurate is the outcomes when an orbital velocity conversion factor based on the linear theory is implemented to convert velocity data to water surface elevation. Because of that, an accurate evaluation of the final results from this method

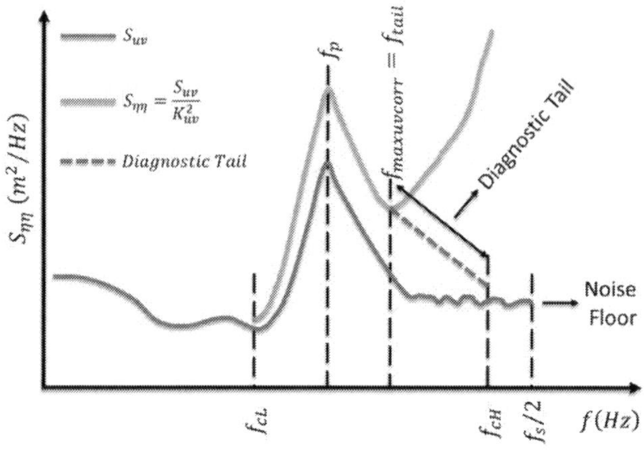

Figure 7.8 Schematic illustration of defining a low-cutoff frequency, f_{cL}, a high-cutoff frequency, f_{cH}, and $f_{maxuvcorr}$. Plot shows a schematic combined horizontal orbital velocities power spectral density, S_{uv}, within a frequency range of $0 \le f \le f_s/2$, and a water surface elevation power spectral density as $S_{\eta\eta} = S_{uv}/K_{uv}^2$ within a frequency range of $f_{cL} \le f \le f_{cH}$. In this plot, all values with a frequency range of $f_{cL} \le f \le f_{cH}$ are retained, and all values with a frequency range of $f < f_{cL}$ or $f > f_{cH}$ are removed. The $f_{maxuvcorr}$ is selected equal to a frequency where a spectrum tail starts to rise to infinity. The f_{tail} is chosen equal to $f_{maxuvcorr}$. A spectrum diagnostic tail is used for a frequency range of $f_{maxuvcorr} \le f \le f_{cH}$. All frequency values and a diagnostic tail in the figure are for a demonstration purpose. In practice, these values need to be selected based on the physical properties of the study area, instrumentation, and nature of the research. Similarly, a diagnostic tail might or might not be required.

is required to guarantee validity of the outcomes (See pressure data analysis chapter for more detail).

7.8 Wave spectrum diagnostic tail

There might be situations that a high-frequency region of the measured power spectral density is missing or is not reliable. In other words, a tail section of the spectrum is missing or is not reliable. For example, such a condition can occur if velocity data are collected with a low sampling frequency. Another situation, which most likely causes a problem with a spectrum tail, can occur when data are collected by a velocity probe deployed relatively deep under a water surface compared with a wavelength of the surface wave. It is explained that, when a velocity probe is used to measure wave activities, shorter wave effects may not reach a depth that a velocity probe is located. In such cases, the high-frequency waves may completely dissipate and diminish in the water column before reaching a depth that a velocity probe is located. This makes a velocity probe unable to detect orbital motions of those high-frequency waves, i.e. short waves. In this case, if there is any measurement within that high-frequency range, it is most likely noise which has an insignificant signal-to-noise ratio. In any of these cases, all those high-frequency energies are lost and need to be replaced to avoid an under-estimation of the wave energy calculated from a water surface elevation power spectral density.

Regardless of the reason behind an unreliable spectrum tail, after a water surface elevation power spectral density is calculated from orbital velocity data, a high-frequency region of the water surface elevation spectrum need to be treated to avoid error in data analysis, particularly to prevent an under-estimation of the wave energy calculated from a water surface elevation spectrum. For example, after a wave spectrum is calculated from orbital velocity data, to compensate for the waves that are not detected by a velocity probe in the frequency domain, first, a simple high-cutoff frequency, f_{tail}, or a low-pass filter, is defined. The f_{tail} defines a frequency where spectrum data with $f > f_{tail}$ are unreliable. This high-cutoff frequency is used to remove a high-frequency part of the spectrum, i.e. to remove spectrum values for a frequency range of $f > f_{tail}$, which are most likely noises. After all noises are screened out, the lost high-frequency energies which have a frequency range of $f > f_{tail}$ need to be replaced. To do that an empirical spectrum tail, a so-called diagnostic tail, is used for a frequency range of $f > f_{tail}$ (e.g. Smith, 2002; Jones and Monismith, 2007; Siadatmousavi et al. 2012).

Depending on the measurement environment, a proper spectrum tail from an empirical wave spectrum is used to replace a tail of the measured water surface elevation spectrum. Commonly, one of the two well-known spectra, i.e. JONSWAP spectrum (Hasselmann et al., 1973) or TMA spectrum (Bouws *et al.*, 1985), is used for this purpose. For data collected in deep water, a noisy section of the spectrum tail is usually replaced with the JONSWAP spectrum tail (Hasselmann et al., 1973), while for data collected in intermediate and shallow water, a noisy section of the spectrum tail is commonly replaced with the TMA spectrum tail (Bouws *et al.*, 1985). A high-frequency section of the water surface elevation power spectral density is replaced by the JONSWAP diagnostic tail proportional to f^{-n} as (e.g. Siadatmousavi et al. 2012):

$$S_{\eta\eta}(f) = S_{\eta\eta}(f_{tail}) \times \left(\frac{f}{f_{tail}}\right)^{-n} \qquad \text{for } f > f_{tail} \tag{7.27}$$

Similarly, a high-frequency section of the water surface elevation power spectral density is replaced by the TMA spectrum tail proportional to f^{-n} as:

$$S_{\eta\eta}(f) = \frac{S_{\eta\eta}(f_{tail})}{\Phi(f_{tail},h)} \times \Phi(f,h) \times \left(\frac{f}{f_{tail}}\right)^{-n} \qquad \text{for } f > f_{tail} \tag{7.28}$$

where, $S_{\eta\eta}(f_{tail})$ is a value of the measured spectrum, $S_{\eta\eta}$, at the frequency f_{tail}, the Φ is a transformation function, and n shows a tail power.

The f_{tail} is a frequency where all values of the measured spectrum with $f > f_{tail}$ are replaced by an empirical spectrum diagnostic tail. There are recommendations to set a value of f_{tail} equal to $f_{tail} = 2.5 f_m$ (Ardhuin et al. 2010) or equal to $f_{tail} = max(2.5 f_m\,, 4 f_{mPM})$ (Siadatmousavi et al. 2012), where $f_m = 1/T_{m01}$ is a mean wave frequency of the measured spectrum, f_{mPM} is a mean wave frequency of the Pierson-Moskowitz spectrum (Pierson and Moskowitz, 1964), $T_{m01} = m_0/m_1$ is a mean wave period, and $m_n = \int_0^\infty S_{\eta\eta}(f) f^n df$ is the nth moment of the wave energy spectrum, e.g. m_0 is a zero-moment of the wave energy spectrum. However, these recommendations might not be suitable for data that are collected in intermediate and shallow water and might result in an inadequate value for f_{tail} in depth-limited conditions. Therefore, for such depth-limited conditions, an adequate value for f_{tail} that reflects this measurement environment and condition needs to be selected. For a water surface elevation spectrum obtained from orbital velocity data, f_{tail} is recommended to be set equal to $f_{tail} = f_{maxuvcorr}$, because data with a frequency range of $f_{maxuvcorr} < f \leq f_{cH}$ might not be reliable. For any given condition, a visual evaluation of the power spectral density might be required to assure a chosen value for f_{tail} is acceptable.

The n in f^{-n} shows a tail power coefficient and defines a slope of the spectrum tail. A value of n depends on physical properties of the water body that data are collected in that. However, typically, a value of n is set equal to 5 for deep and 3 for shallow water (e.g. Phillips, 1958; Thornton, 1977; Kitaigorodskii, 1983; Miller and Vincent, 1990; Smith, 2002; Jones and Monismith, 2007; Holthuijsen, 2007; Kaihatu et al. 2007; Siadatmousavi et al. 2012). Therefore, a value of $n = 4$ might be suitable for intermediate depth.

The Φ is a transformation function. It transforms the JONSWAP spectrum in deep water into the TMA spectrum in depth-limited water as (Kitaigordskii et al., 1975; Hughes, 1984; Bouws et al., 1985; Holthuijsen, 2007):

$$\Phi = \frac{R_{\omega_h}^{-2}}{1+2\omega_h^2 R_{\omega_h}/\sinh\left(2\omega_h^2 R_{\omega_h}\right)} \tag{7.29}$$

where, R_{ω_h} and ω_h are:

$$R_{\omega_h} \tanh\left(\omega_h^2 R_{\omega_h}\right) = 1 \tag{7.30}$$

$$\omega_h = 2\pi f \sqrt{\frac{h}{g}} \tag{7.31}$$

A value of Φ can be calculated from (Tucker, 1994; Holthuijsen, 2007):

$$\Phi = \frac{\tanh^2(kh)}{1+2kh/\sinh(2kh)} \tag{7.32}$$

where, k is a wavenumber and h is water depth. Also, a value of Φ can be approximated by (Thompson and Vincent, 1983; Bergdahl, 2009):

$$\Phi(f,h) \approx \frac{\omega_h^2}{2} \quad \text{for} \quad \omega_h \leq 1 \tag{7.33}$$

$$\Phi(f,h) \approx 1 - 0.5(2-\omega_h)^2 \quad \text{for} \quad 1 < \omega_h < 2 \tag{7.34}$$

$$\Phi(f,h) = 1 \quad \text{for} \quad \omega_h \geq 2 \tag{7.35}$$

In general, it is not recommended to implement a diagnostic tail to replace a spectrum tail of the measured data. However, there might be situations that a tail of the measured spectrum is required to be replaced. For example, such situations that a high-frequency data are missing might occur if data are collected with a low sampling frequency, or data are collected with a velocity probe that is deployed too deep in the water column. Also, if the high-frequency data are contaminated by noise, it needs to be replaced by a diagnostic tail (See pressure data analysis chapter for more detail).

8 Calculating Secondary Wave Properties

In this chapter, calculations of the secondary wave properties besides wave height and a wave period are covered. First, calculations of the wavenumber and wavelength are explained. Next, relationships between wave properties in the time and frequency domains are discussed. Then, calculations of the wave properties from the linear wave theory are described. After that, wave height and wave period distributions are explained. Next, directional wave spectra are discussed. Step by step method to generate a time series from a power spectral density is presented. Methods to separate wind sea and swell waves are explained.

8.1 Calculating wavenumber and wavelength

A wavelength, L, and a wavenumber, $k = 2\pi/L$, are required for a wide range of wave data analysis calculations. Values of k and L, can be obtained by solving a linear wave dispersion relationship, i.e. $\omega^2 = gk\tanh(kh)$, numerically, where $\omega = 2\pi/T = 2\pi f = Ck$ is a wave angular frequency, h is water depth, and C is a wave celerity. It should be pointed out that a dispersion relationship cannot be solved explicitly. In case that a bottom current with a mean flow velocity of U is present in the area, then the

Doppler-shift effect needs to be considered in a dispersion relationship as $\omega^2 = gk\tanh(kh) + kU$ (e.g. Smith, 2002; Jones and Monismith 2007).

As pointed out, a dispersion relationship, $\omega^2 = gk\tanh(kh)$, cannot be resolved explicitly. It either can be solved numerically or can be approximated by proposed explicit relationships in the literature. One method to obtain values of the wavelength and wavenumber is to solve a dispersion relationship using a solver function provided by programming languages. A wavenumber and a wavelength are calculated by using MATLAB or GNU Octave as:

```
%Input parameters
h=1; %Water depth in (m)
T=3; %Wave period in (s)

%Initial value for k
w=2*pi/T; %Wave angular frequency, w=2*pi/T
k0=(w^2)/9.81; %Deep water wavenumber
k0h=k0*h;
kini=k0h/h; %Initial value for k

%Calculating a wavenumber (k) and a wavelength (L)
fun = @(x)(w^2-(9.81*x*tanh(x*h))); %w^2=g*k*tanh(kh)
k=fzero(fun,kini); %Wavenumber
L=2*pi/k; %Wavelength
```

Here is a Python code for calculating a wavenumber and a wavelength. Note that you need to install and import NumPy package as np and SciPy package as sp along with Python.

```
#Importing required packages
import numpy as np
import scipy as sp
from scipy import optimize

#Input parameters
h=1 #Water depth in (m)
T=3 #Wave period in (s)

#Initial value for k
w=2*np.pi/T #Wave angular frequency, w=2*pi/T
k0=(w**2)/9.81 #Deep water wavenumber
k0h=k0*h
kini=k0h/h #Initial value for k
```

```
#Calculating a wavenumber (k) and wavelength (L)
fun = lambda x : w**2-(9.81*x*np.tanh(x*h)) #w**2=g*k*tanh(kh)
k=sp.optimize.fsolve(fun,kini) #Wavenumber
L=2*np.pi/k #Wavelength
```

In case that a solver function is not available, a method suggested by Goda (2010) can be used as an alternative method to solve a dispersion relationship. Goda (2010) suggested that a wavenumber and a wavelength can be estimated by re-writing a dispersion relationship as $kh \tanh(kh) = k_0 h$ or $kh - k_0 h \coth(kh) = 0$ and solving it iteratively based on Newton's method as (Goda, 2010; Reeve et al., 2012):

$$(kh)_2 = (kh)_1 - \frac{(kh)_1 - (k_0 h)\coth(kh)_1}{1 + (k_0 h)(\coth^2(kh)_1 - 1)} \tag{8.1}$$

where, h is water depth, k is a wavenumber, $k_0 = \omega^2/g$ is a deep-water wavenumber, g is the gravitational acceleration, and $\omega = 2\pi/T = 2\pi f$ is a wave angular frequency. A value of the kh can be calculated iteratively by using $(kh)_2$ from current iteration as $(kh)_1$ in the next iteration, until $(kh)_1$ and $(kh)_2$ become close enough. An initial estimation for $(kh)_1$ for the first iteration is:

$$(kh)_1 = \begin{cases} k_0 h, & k_0 h \geq 1 \\ (k_0 h)^{0.5}, & k_0 h < 1 \end{cases} \tag{8.2}$$

In this method, an absolute error, which is calculated as $|1 - (kh)_2/(kh)_1|$, is reported to be less than 0.05% after 3 iterations (Goda, 2010; Reeve et al., 2012).

Goda (2010) method can be implemented by a programming language to solve a dispersion relationship, $\omega^2 = gk\tanh(kh)$, and obtain a wavenumber and a wavelength. Here is a MATLAB or GNU Octave code for calculating a wavenumber and a wavelength using Goda (2010) method with 3 iterations:

```
%Input parameters
h=1; %Water depth in (m)
T=3; %Wave period in (s)

%Initial value for kh
```

```
w=2*pi/T; %Wave angular frequency, w=2*pi/T
k0=(w^2)/9.81; %Deep water wavenumber
k0h=k0*h;
if k0h>=1
    kh=k0h;
elseif k0h<1
    kh=k0h^0.5;
end

%Calculating a wavenumber (k) and a wavelength (L)
for i=1:3
    kh=kh-((kh-k0h.*coth(kh))./(1+k0h.*((coth(kh)).^2-1)));
%Calculating kh
end
k=kh/h; %Wavenumber
L=2*pi/k; %Wavelength
```

Here is a Python code for calculating a wavenumber and a wavelength using Goda (2010) method with 3 iterations. Note that you need to install and import NumPy package as np along with the Python.

```
#Importing required packages
import numpy as np

#Input parameters
h=1 #Water depth in (m)
T=3 #Wave period in (s)

#Initial value for k
w=2*np.pi/T #Wave angular frequency, w=2*pi/T
k0=(w**2)/9.81 #Deep water wavenumber
k0h=k0*h
if k0h>=1:
    kh=k0h
elif k0h<1:
    kh=k0h**0.5

#Calculating a wavenumber (k) and a wavelength (L)
for i in range(0,3,1):
    kh=kh-((kh-k0h*(np.tanh(kh))**-1)/(1+k0h*((np.tanh(kh))**(-
2)-1))); #Calculating kh
k=kh/h #Wavenumber
L=2*np.pi/k #Wavelength
```

There are situations that exact values of the wavenumber and wavelength might not be required or might not be feasible. For example, using a numerical method to calculate exact wavenumber and wavelength values for a big dataset can become numerically expensive. In such conditions, using accurate explicit estimations of the wavenumber and wavelength are acceptable in most cases. There have been a number of methods developed to estimate a wavenumber and a wavelength explicitly with an acceptable accuracy (e.g. Hunt, 1979; Fenton and McKee, 1990; Guo, 2002; Soulsby, 2006; Yamaguchi and Nonaka, 2007; You, 2008; Beji, 2013; Vatankhah and Aghashariatmadari, 2013; Yu, 2013; Yamaguchi et al., 2014). One of the widely used equations for direct wavenumber estimation was proposed by Hunt (1979) in the Pade approximant from as:

$$\frac{c^2}{gh} = (k_0h + (1 + 0.6522(k_0h) + 0.4622(k_0h)^2 + 0.0864(k_0h)^4 + 0.0675(k_0h)^5)^{-1})^{-1} \tag{8.3}$$

where,

$$k_0h = \frac{\omega^2}{g}h \tag{8.4}$$

$$k = \frac{\omega}{c} \tag{8.5}$$

Also, Hunt (1979) equation in another form is presented as:

$$(kh)^2 = (k_0h)^2 + \frac{k_0h}{1+\sum_{i=1}^{\infty} d_i(k_0h)^i} \tag{8.6}$$

where, $k_0 = \omega^2/g$ is a deep water wavenumber, g is the gravitational acceleration, $d_1 = 0.6666666667$, $d_2 = 0.3555555556$, $d_3 = 0.1608465608$, $d_4 = 0.0632098765$, $d_5 = 0.0217540484$, and $d_6 = 0.0065407983$ (Hunt 1979; You, 2008).

Beji (2013) proposed a more accurate equation to estimate wavenumber explicitly with a maximum absolute relative error of 0.044% as:

$$kh = \frac{k_0h\left(1+(k_0h)^{1.09}\exp\left(-(1.55+1.3(k_0h)+0.216(k_0h)^2)\right)\right)}{\sqrt{\tanh(k_0h)}} \tag{8.7}$$

Vatankhah and Aghashariatmadari (2013) proposed a modified Beji (2013) equation for an explicit estimation of the wavenumber with a maximum absolute relative error of 0.001% as:

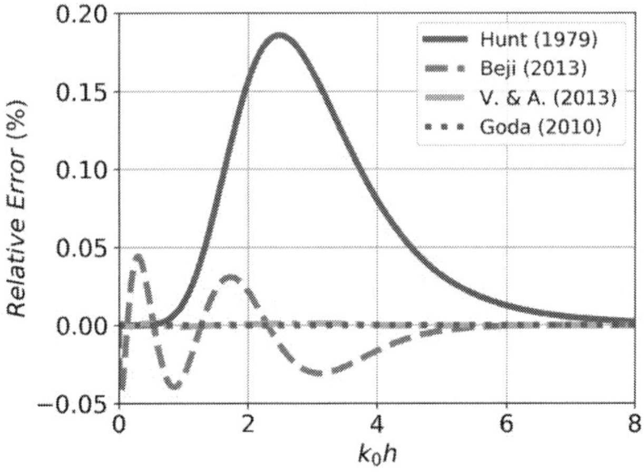

Figure 8.1 Comparison of a relative error in estimations of the wavenumber, k, from Hunt (1979), Beji (2013), Vatankhah and Aghashariatmadari (2013), and Goda (2010). Values of k are estimated for a water depth of $h = 10\ m$ and k_0h range of $0 \leq k_0h \leq 8$. Results for Goda (2010) are reported after 3 iterations. A relative error, ε, is calculated in percent as $\varepsilon = (k_{est} - k_{exa})/k_{exa} \times 100$ or $\varepsilon = (k_{est}/k_{exa} - 1) \times 100$, where ε is a relative error, k_{exa} is an exact wavenumber, and k_{est} is an estimated wavenumber. V. & A. (2013) refers to Vatankhah and Aghashariatmadari (2013)

$$kh = \frac{k_0h + (k_0h)^2 \exp\left(-\left(3.2 + (k_0h)^{1.65}\right)\right)}{\sqrt{\tanh(k_0h)}} + k_0h\left(1 - \exp\left(-(k_0h)^{0.132}\right)\right)^{\left(5.0532 + 2.158(k_0h)^{1.505}\right)} \tag{8.8}$$

Note that, to use a numerical solver to obtain an exact wavenumber value from a dispersion relationship, a solver requires an initial estimation of the wavenumber. In such cases, an estimated wavenumber from one of these explicit methods such as Hunt (1979) or Beji (2013) can be used as an initial estimation in a numerical solution of the dispersion relationship.

Finally, after a wavenumber, k, is calculated, a wavelength, L, is obtained as:

$$L = \frac{2\pi}{k} \tag{8.9}$$

8.2 Relationship between wave properties in the time and frequency domains

In practice, researchers study the properties of the various time series datasets. Wave properties of these datasets can be calculated either in the time domain or in the frequency domain. In the time domain, a zero-crossing method is used to calculate wave properties. In the frequency domain, the spectral analysis is used to calculate wave properties. Sometimes, it is required to convert a calculated value from one of these two methods to an equivalent one from the other method. For example, it may be required to convert significant wave height, H_s, which is calculated in the time domain, to zero-moment wave height, H_{m0}, which is an equivalent wave height representation in the frequency domain. Relationships between wave properties estimated in the time domain with equivalent ones in the frequency domain are (e.g. Holthuijsen, 2007):

$$H_s \approx H_{m0} \tag{8.10}$$

$$H_{rms} \approx \frac{H_{m0}}{\sqrt{2}} \tag{8.11}$$

$$\eta_{rms} = \sqrt{m_0} \tag{8.12}$$

$$T_z \approx T_{m02} \tag{8.13}$$

$$T_s \approx 0.95T_p \tag{8.14}$$

where, H_s is significant wave height, $H_{m0} = 4\sqrt{m_0}$ is zero-moment wave height, H_{rms} is the root mean square (rms) wave height, η_{rms} is a root mean square of the water surface elevation, T_z is a zero-crossing mean wave period, T_s is a significant wave period, $T_{m02} = \sqrt{m_0/m_2}$ is a mean zero-crossing wave period, $T_p = 1/f_p$ is a peak wave period, $m_0 = \int_0^\infty S_{\eta\eta}(f)df$ is the zero-moment of the water surface elevation power spectral density, $m_n = \int_0^\infty S_{\eta\eta}(f)f^n df$ is the nth moment of the water surface elevation power spectral density, f is frequency, f_p is a peak wave frequency, and $S_{\eta\eta}$ is a water surface elevation power spectral density. Note that Holthuijsen (2007) suggested $H_s \approx 0.927H_{m0}$.

8.3 Calculating wave properties from the linear wave theory

Wave properties, particularly wave height and a wave period, are mostly obtained either from a zero-crossing method in the time domain or from the spectral analysis in the frequency domain. These estimated wave properties in either of time or frequency domain are mostly statistical representations of these waves. In other words, values from either zero-crossing or spectral analysis method are representatives of the measured wave field. For example, significant wave height, H_s, represents a mean value of the top third wave heights in a dataset. Using these statistical approaches to evaluate a wave dataset is based on the assumption that the real-world wave fields are generally generated by superposition of a series of random waves. Therefore, statistical values are used to describe them.

After wave height and a wave period are calculated either in the time or frequency domain, rest of wave properties can be obtained from these two parameters. Some of these wave properties that may be obtained from wave height and a wave period are a wavenumber, a wavelength, a wave celerity, wave energy, wave power, wave radiation stress, wave orbital velocity, wave orbital excursion, wave-induced current, and wave bottom friction. However, as mentioned above, wave height and a wave period calculated from the zero-crossing or spectral analysis are statistical representations of the random waves. On the other hand, calculation of other wave properties from wave height and a wave period relies on the linear wave theory, which is developed for a simple sinusoidal wave and is not suitable for random waves. Because of that, wave height and a wave period estimated by the zero-crossing or spectral analysis may not be used directly in the linear wave theory equations to estimate other wave properties such as a wavenumber, a wavelength, and a wave celerity.

To address this issue, in practice, first, wave height and a wave period calculated by the zero-crossing or spectral analysis are converted to equivalent wave height and wave period for a linear sinusoidal wave. Then, these equivalent wave height and wave period are used to estimate other wave properties for given water depth. To obtain equivalent wave height, the linear wave height, H, is approximated by $H \approx H_{rms} \approx H_s/\sqrt{2}$ in the time domain and by $H \approx H_{m0}/\sqrt{2}$ in the frequency domain. For example, wave energy, E_w, is estimated in the time domain from significant wave height, H_s, as:

$$E_w = \frac{1}{8}\rho g H^2 \approx \frac{1}{8}\rho g H_{rms}^2 \approx \frac{1}{8}\rho g \left(\frac{H_s}{\sqrt{2}}\right)^2 \quad (8.15)$$

Likewise, wave energy, E_w, is estimated in the frequency domain from zero-moment wave height, H_{m0}, as:

$$E_w = \frac{1}{8}\rho g H^2 \approx \frac{1}{8}\rho g \left(\frac{H_{m0}}{\sqrt{2}}\right)^2 \quad (8.16)$$

or

$$E_w = \rho g \int_0^\infty S_{\eta\eta}(f)df \approx \sum_{i=1}^{N} \frac{1}{8}\rho g \left(\frac{4\sqrt{S_{\eta\eta,i} \times \Delta f_i}}{\sqrt{2}}\right)^2 \approx \frac{1}{8}\rho g \left(\frac{H_{m0}}{\sqrt{2}}\right)^2 \quad (8.17)$$

Similarly, to obtain an equivalent wave period, the linear wave period, T, is approximated by $T \approx T_z$ in the time domain and by $T \approx T_{m02}$ in the frequency domain. For example, wave celerity, C, is calculated as:

$$C = \frac{gT}{2\pi}\tanh(kh) \approx \frac{gT_z}{2\pi}\tanh(kh) \quad (8.18)$$

Also, it is common to use a peak wave period, $T_P = 1/f_P$, instead of the wave period, T, in wave property calculations. In such cases, all estimated wave properties represent properties of the peak wave. For example, a peak wave celerity, C_p, is calculated as:

$$C_p = \frac{gT_p}{2\pi}\tanh(k_p h) \quad (8.19)$$

where, k_p is a peak wavenumber, which is calculated from a linear wave dispersion relationship for a peak period, i.e. it is calculated from $(2\pi/T_p)^2 = gk_p\tanh(k_p h)$.

It should be pointed out that, using the linear wave theory in areas that have dominant non-linear and irregular waves can results in error in wave parameter estimations.

8.4 Wave height and wave period distributions

It is accepted that if wave heights are narrow banded, i.e. if wave periods (frequencies) have a small range, then wave heights can be well approximated by the Rayleigh distribution. A probability density function, f_{PDF}, of the wave height following the Rayleigh distribution is written as (e.g. Longuet-Higgins, 1952; Young, 1999; Dean and Dalrymple, 1991; Sorensen, 2005; Holthuijsen, 2007; McCormick, 2009; Casas-Prat and Holthuijsen, 2010):

$$f_{pdf}(H) = \left(\frac{2H}{H_{rms}^2}\right) \exp\left(-\left(\frac{H}{H_{rms}}\right)^2\right) \tag{8.20}$$

or

$$f_{pdf}(H) = \left(\frac{H}{4m_0}\right) \exp\left(-\frac{H^2}{8m_0}\right) \tag{8.21}$$

or

$$f_{pdf}\left(\frac{H}{\sqrt{m_0}}\right) = \frac{1}{4}\left(\frac{H}{\sqrt{m_0}}\right) \exp\left(-\frac{1}{8}\left(\frac{H}{\sqrt{m_0}}\right)^2\right) \tag{8.22}$$

where, $f_{pdf}(H)$ is a probability density function of the wave height, H is wave height, $H_{rms} \approx H_{m0}/\sqrt{2}$ is the root mean square (rms) wave height, $H_{m0} = 4\sqrt{m_0}$ is zero-moment wave height, $m_0 = \int_0^\infty S_{\eta\eta}(f)df$ is the zero-moment of the water surface elevation power spectral density, $S_{\eta\eta}$ is a water surface elevation power spectral density, and f is frequency. Note that the total area under a probability density function is unity, i.e. $\int_0^\infty f_{pdf}(H)\, dH = 1$.

For a probability density function of the wave period, Bretschneider (1959) proposed a probability density function as (e.g., Bretschneider, 1959; Chakrabarti, 1987):

$$f_{pdf}(T) = 2.7\frac{T^3}{\bar{T}^4}\exp\left(-0.675\left(\frac{T}{\bar{T}}\right)^4\right) \tag{8.23}$$

where, $f_{pdf}(T)$ is a probability density function of the wave period, T is a wave period, and $\bar{T}$ is a mean wave period.

Later, Longuet-Higgins (1975) and Longuet-Higgins (1983) proposed a wave period probability density function for a narrow spectrum (e.g. Longuet-Higgins, 1975; Longuet-Higgins, 1983; Liu et al., 1997; Soares and Carvalho, 2003; Xu et al., 2004; Kim, 2010; Zhang et al., 2013). Longuet-Higgins (1975) suggested a probability density function as:

$$f_{pdf}(T) = \frac{\nu^2\bar{T}^2}{2(\nu^2\bar{T}^2+(T-\bar{T})^2)^{3/2}} \tag{8.24}$$

or

$$f_{pdf}(\tau) = \frac{\nu^2}{2(\nu^2+(\tau-1)^2)^{3/2}} \tag{8.25}$$

Longuet-Higgins (1983) suggested a probability density function as:

$$f_{pdf}(\tau) = \frac{L_\nu}{2\nu\tau^2}\left(1+\frac{(1-1/\tau)^2}{\nu^2}\right)^{-3/2} \tag{8.26}$$

where, $f_{pdf}(\tau)$ is a probability density function of the dimensionless wave period, $\tau = T/\bar{T}$ is a dimensionless wave period, $\bar{T} = T_{m01} = m_0/m_1$ is a mean wave period, $\nu = \sqrt{(m_0 m_2 - m_1^2)/m_1^2}$ is a spectral bandwidth, $L_\nu = 2/(1+(1+\nu^2)^{-0.5}) \approx 1+\frac{1}{4}\nu^2$ is a normalization factor which makes a probability density function equal to unity (e.g. Longuet-Higgins 1983; Zhang et al., 2013), and $m_n = \int_0^\infty S_{\eta\eta}(f)f^n df$ is the nth moment of the water surface elevation power spectral density.

For a joint probability distribution of the wave amplitude and wave period, Longuet-Higgins (1983) proposed a probability density function as (e.g., Longuet-Higgins 1983; Xu et al., 2004; Kim, 2010; Zhang et al., 2013):

$$f_{pdf}(R,\tau) = \frac{2L_\nu R^2}{\sqrt{\pi}\nu\tau^2}\times\exp\left(-R^2\left(1+\frac{(1-1/\tau)^2}{\nu^2}\right)\right) \tag{8.27}$$

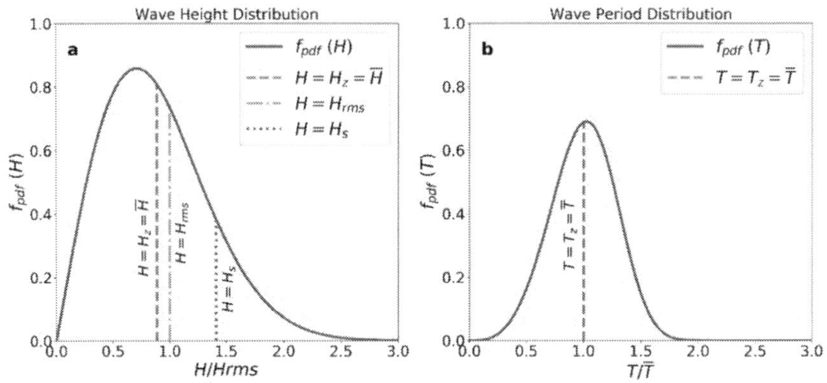

Figure 8.2 Distribution of the wave height (a) and wave period (b). A wave height Rayleigh distribution, $f_{pdf}(H)$, is plotted for $0 \leq H \leq 3\ m$ and $H_{rms} = 1\ m$. A wave period distribution, $f_{pdf}(T)$, is plotted for $0 \leq T \leq 6\ s$ and $\bar{T} = 2\ s$, based on Bretschneider (1959). In a plot (a), the first vertical line from the left represents $H = H_z = \bar{H}$ or $H/H_{rms} = H_z/H_{rms} \approx 0.866$, the second vertical line from the left represents $H = H_{rms}$ or $H/H_{rms} = H_{rms}/H_{rms} = 1$, and the third vertical line from the left represents $H = H_s$ or $H/H_{rms} = H_s/H_{rms} \approx \sqrt{2}$. In a plot (b), the vertical line represents $T = T_z = \bar{T}$.

where, $f_{pdf}(R, \tau)$ is a joint probability density function for a dimensionless wave amplitude and wave period, $R = a/\sqrt{2m_0}$ is a dimensionless wave amplitude, a is a wave amplitude, and m_0 is the zero-moment of the water surface elevation power spectral density.

In experimental studies, it is common to calculate a probability density distribution of the measured data and compare it with theoretical or empirical ones. To calculate a probability density distribution of the measured data, first, a histogram of the dataset is calculated. Then, these histogram values are divided by the total area of the histogram to attain a probability density distribution. To obtain the total area of the histogram, first, a value of histogram at each bin is multiplied by a bin width of that bin, and then, results for all bins are added together.

8.5 Directional wave spectra

In the real world, waves propagate in different directions which results in wave energy distribution in different directions. This energy distribution is typically presented by a directional spectrum. To define a directional spectrum, a mean direction of the wave propagations and a distribution of the wave propagations are required. If these values are known, a directional water surface elevation power spectral density, $S_{\eta\eta}(f,\theta)$, is expressed as:

$$S_{\eta\eta}(f,\theta) = S_{\eta\eta}(f)D(f,\theta) \tag{8.28}$$

where, $S_{\eta\eta}(f,\theta)$ is a directional water surface elevation power spectral density, $S_{\eta\eta}(f)$ is a non-directional water surface elevation power spectral density, $D(f,\theta)$ is a directional spreading function, which shows a directional distribution of the wave propagation, f is a frequency, and θ is a wave direction. A directional spreading function should follow $\int_{-\pi}^{\pi} D(f,\theta)d\theta = 1$, therefore $S_{\eta\eta}(f) = \int_{-\pi}^{\pi} S_{\eta\eta}(f,\theta)d\theta$ and $\int_0^{\infty}\int_{-\pi}^{\pi} S_{\eta\eta}(f)D(f,\theta)df d\theta = \int_0^{\infty} S_{\eta\eta}(f)df = m_0$.

As pointed out, to define a directional spectrum, a mean wave direction and a directional spreading of the wave propagation around a mean direction are required. A directional spreading function either can be calculated directly from measured data or can be estimated using parametric relationships. One of the earliest parametric directional spreading functions is (St. Denis and Pierson, 1953; Pierson et al., 1955; Memos and Ziros, 2003; Sorensen, 2005):

$$D(f,\theta) = \begin{cases} \frac{2}{\pi}\cos^2(\theta), & |\theta - \theta_m| \le \frac{\pi}{2} \\ 0, & |\theta - \theta_m| > \frac{\pi}{2} \end{cases} \tag{8.29}$$

Longuet-Higgins et al. (1963) proposed a parametric directional spreading function as (Goda, 1999; Sorensen, 2005):

$$D(f,\theta) = G(s)\cos^{2s}\left(\frac{\theta-\theta_m}{2}\right) \quad \text{for} \quad -\pi \le \theta \le \pi \tag{8.30}$$

$$G(s) = \left(\frac{2^{2s-1}}{\pi}\right)\left(\frac{\Gamma^2(s+1)}{\Gamma(2s+1)}\right) \tag{8.31}$$

where, θ_m is a mean wave direction, s is a spreading parameter which represents a directional spreading of the wave energy around θ_m (Deo et al., 2002), and Γ is the Gamma function. To use this relationship, values of s and θ_m needs to be defined first. Mitsuyasu et al. (1975) suggested that s can be estimated as (Ewans, 1998; Goda, 1999; Hwang and Wang, 2001):

$$s = \begin{cases} s_p \left(\frac{f}{f_p}\right)^5, & f < f_p \\ s = s_p \left(\frac{f}{f_p}\right)^{-2.5}, & f \geq f_p \end{cases} \tag{8.32}$$

$$s_p = 11.5 \left(\frac{U_{10}}{C_p}\right)^{-2.5} \tag{8.33}$$

where, f_p is a peak wave frequency, U_{10} is a wind velocity at height $10\,m$ above a surface, $C_p = g/(2\pi f_p)$ is a peak wave phase speed (a peak wave celerity) in deep water, which is associated with a peak wave frequency, and s_p is a spreading parameter associated with a peak wave frequency.

Later, Hasselmann et al. (1980) suggested that s can be estimated as (Ewans, 1998; Hwang and Wang, 2001):

$$s = \begin{cases} 6.97 \left(\frac{f}{f_p}\right)^{4.06}, & f < 1.05 f_p \\ s = 9.77 \left(\frac{f}{f_p}\right)^{\mu}, & f \geq 1.05 f_p \end{cases} \tag{8.34}$$

$$\mu = -2.33 - 1.45 \left(\frac{U_{10}}{C_p} - 1.17\right) \tag{8.35}$$

Donelan et al. (1985) proposed a directional spreading function as:

$$D(f,\theta) = 0.5\beta \operatorname{sech}^2\left(\beta(\theta - \theta_m)\right) \qquad \text{for} \qquad -\pi \leq \theta \leq \pi \tag{8.36}$$

where, β is a spreading parameter and can be estimated as (Donelan et al., 1985; Banner, 1990; Ewans, 1998; Massel and Brinkman, 1998; Goda, 1999; Hwang and Wang, 2001):

$$\beta = \begin{cases} 1.24, & \frac{f}{f_p} \leq 0.56 \\ \beta = 2.61\left(\frac{f}{f_p}\right)^{1.3}, & 0.56 < \frac{f}{f_p} \leq 0.95 \\ \beta = 2.28\left(\frac{f}{f_p}\right)^{-1.3}, & 0.95 < \frac{f}{f_p} \leq 1.6 \\ \beta = 10^{\left(-0.4+0.8393\exp\left(-0.567\ln\left(\left(\frac{f}{f_p}\right)^2\right)\right)\right)}, & \frac{f}{f_p} > 1.6 \end{cases} \quad (8.37)$$

Initially, Donelan et al. (1985) suggested $\beta = 1.24$ for $f/f_p > 1.6$, but Banner (1990) proposed a non-constant value for β as $\beta = 10^{\left(-0.4+0.8393\exp\left(-0.567\ln\left(\left(\frac{f}{f_p}\right)^2\right)\right)\right)}$ for $f/f_p > 1.6$.

As mentioned above, to define a directional spectrum, a directional spreading function and a mean wave direction of the wave propagations are required. A mean wave direction is typically defined in the frequency domain. To do that, first, a directional spreading function is written in the form of the Fourier series as (Ewans, 1998; Pedersen et al., 2005):

$$D(f,\theta) = \frac{1}{\pi}\left(\frac{1}{2} + \sum_n (a_n(f)\cos(n\theta) + b_n(f)\sin(n\theta))\right) \quad (8.38)$$

From that, a directional spreading function for the first and second order Fourier coefficients would be (Deo et al., 2002):

$$D(f,\theta) = \frac{1}{\pi}\Big(\frac{1}{2} + a_1(f)\cos(\theta) + b_1(f)\sin(\theta) + a_2(f)\cos(2\theta) + b_2(f)\sin(2\theta)\Big) \quad (8.39)$$

where, a_1, b_1, a_2, and b_2 are Fourier coefficients.

Here, it is explained how to define these coefficients for a dataset contains water pressure and wave horizontal orbital velocity. Method to obtain the Fourier coefficients for this type of dataset, a so-called PUV, are (e.g. Herbers et al., 1999):

$$a_1(f) = \frac{C_{\eta u}(f)}{\left(C_{\eta\eta}(f)(C_{uu}(f)+C_{vv}(f))\right)^{0.5}} \quad (8.40)$$

$$b_1(f) = \frac{C_{\eta v}(f)}{\left(C_{\eta\eta}(f)\left(C_{uu}(f)+C_{vv}(f)\right)\right)^{0.5}} \tag{8.41}$$

$$a_2(f) = \frac{C_{uu}(f)-C_{vv}(f)}{C_{uu}(f)+C_{vv}(f)} \tag{8.42}$$

$$b_2(f) = \frac{2C_{uv}(f)}{C_{uu}(f)+C_{vv}(f)} \tag{8.43}$$

$$C_{\eta\eta}(f) = S_{\eta\eta}(f) = \frac{1}{K_p^2(f)} \times \frac{S_{PP}(f)}{\rho^2 g^2} \tag{8.44}$$

$$C_{uu}(f) = S_{uu}(f) = \frac{S_{uu}(f)}{K_{uv}^2(f)} \tag{8.45}$$

$$C_{vv}(f) = S_{vv}(f) = \frac{S_{vv}(f)}{K_{uv}^2(f)} \tag{8.46}$$

$$C_{\eta u}(f) = \frac{1}{K_{uv}(f)K_p(f)} \times \frac{S_{Pu}(f)}{\rho g} \tag{8.47}$$

$$C_{\eta v}(f) = \frac{1}{K_{uv}(f)K_p(f)} \times \frac{S_{Pv}(f)}{\rho g} \tag{8.48}$$

$$C_{uv}(f) = \frac{S_{uv}(f)}{K_{uv}^2(f)} \tag{8.49}$$

$$K_p(f, z = -h_p) = \frac{\cosh k(h+z)}{\cosh(kh)} = \frac{\cosh(kd_p)}{\cosh(kh)} \tag{8.50}$$

$$K_{uv}(f, z = -h_{uv}) = \frac{2\pi f \cosh k(h+z)}{\sinh(kh)} = \frac{2\pi f \cosh(kd_{uv})}{\sinh(kh)} \tag{8.51}$$

where, S_{PP}, $S_{\eta\eta}$, S_{uu}, and S_{vv} are power spectral densities for water dynamic pressure, water surface elevation, water particle horizontal orbital velocity in x direction, and water particle horizontal orbital velocity in y direction, respectively. S_{Pu}, S_{Pv}, $S_{\eta u}$, $S_{\eta v}$, S_{uv} are cross power spectral densities for water dynamic pressure and water particle horizontal orbital velocity in x direction, water dynamic pressure and water particle horizontal orbital velocity in y direction, water surface elevation and water particle horizontal orbital velocity in x direction, water surface elevation and water particle horizontal orbital velocity in y direction, water particle horizontal orbital velocity in x and water particle horizontal orbital velocity in y direction, respectively. The x and y are orthogonal horizontal directions (axes). The ρ is the water density, g is the gravitational acceleration, η is a water surface elevation, and P is water dynamic pressure. The K_p is a pressure response factor which represents a dynamic pressure to surface elevation conversion factor. The K_{uv} is a water particle horizontal orbital velocity to the water surface elevation conversion factor. The h is local water depth, z is an upward vertical axis with zero at a water surface, h_p is pressure sensor depth which is equal to $z = -h_p$, and h_{uv} is velocity probe depth which is equal to $z = -h_{uv}$. The d_p is a pressure sensor location distance from a seabed, and d_{uv} is a velocity probe location distance from a seabed. The $f = 1/T$ is a wave frequency, k is a wavenumber, and T is a wave period. A wavenumber is a function of the wave frequency, $f = 1/T$, and local water depth, h. A wavenumber is defined from a water wave dispersion relationship, i.e. $\omega^2 = (2\pi f)^2 = gk\tanh(kh)$, where $\omega = 2\pi f = 2\pi/T$ is an angular frequency.

For other types of dataset such as the ones that contain water surface elevation and wave horizontal orbital velocities, or the ones that contain three components of wave orbital velocities, Fourier coefficients are slightly different (Grosskopf et al., 1983; Earle et al., 1995; Benoit et al., 1997; Herbers et al., 1999; Deo et al., 2002, Kumar and Anand, 2004; and Pedersen et al., 2005).

After Fourier coefficients, i.e. a_1, b_1, a_2, and b_2, are defined, then, a mean wave direction, θ_1, and a principal wave direction, θ_2, as a function of frequency are calculated as (Longuet-Higgins et al., 1963; Ewans, 1998; Earle et al., 1999; Kumar and Anand, 2004):

$$\theta_1(f) = \operatorname{atan2}\left(\frac{b_1(f)}{a_1(f)}\right) \tag{8.52}$$

$$\theta_2(f) = 0.5\operatorname{atan2}\left(\frac{b_2(f)}{a_2(f)}\right) \tag{8.53}$$

where, θ_1 is a mean wave direction, and θ_2 is a principal wave direction.

Another method to calculate a mean wave direction for a given frequency, f_i, is (Work, 2008):

$$\theta_{mean}(f_i) = \tan^{-1}\left(\frac{\sum_j \sin\theta_j S_{\eta\eta}(f_i,\theta_j)}{\sum_j \cos\theta_j S_{\eta\eta}(f_i,\theta_j)}\right) \tag{8.54}$$

where, θ_{mean} is a mean wave direction for a given frequency.

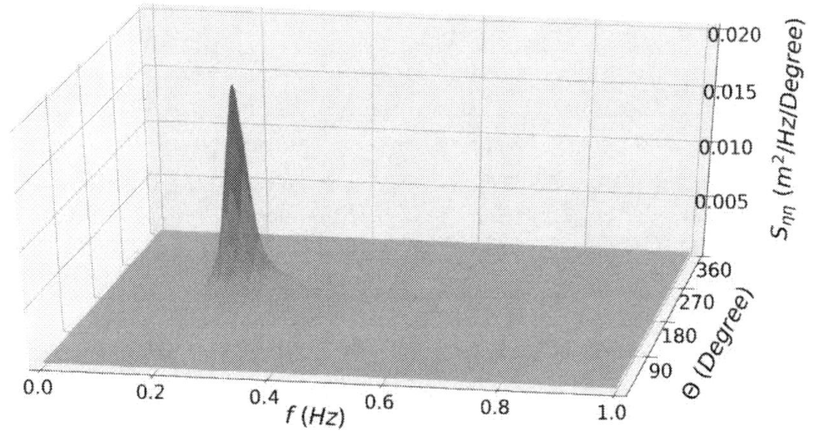

Figure 8.3 Schematic illustration of the water surface elevation directional power spectral density, $S_{\eta\eta}(f,\theta)$. A directional spectrum is calculated from Donelan et al. (1985) for a wind direction of 180°. A sampling frequency is $f_s = 2\ Hz$. A schematic power spectral density is obtained from the JONSWAP spectrum (Hasselmann et al., 1973) for a peak wave frequency of 0.25 Hz.

After a mean wave direction is calculated for each frequency, then a weighted mean direction for the entire spectrum is calculated as (e.g., Work, 2008):

$$\theta_m = \tan^{-1}\left(\frac{\int_0^{f_s/2} \sin\theta_1(f) S_{\eta\eta}(f) df}{\int_0^{f_s/2} \cos\theta_1(f) S_{\eta\eta}(f) df}\right) \approx \tan^{-1}\left(\frac{\sum_i \sin\theta_1(f_i) S_{\eta\eta}(f_i)}{\sum_i \cos\theta_1(f_i) S_{\eta\eta}(f_i)}\right) \tag{8.55}$$

or

$$\theta_m = \tan^{-1}\left(\frac{\sum_i \sin\theta_{mean}(f_i) S_{\eta\eta}(f_i)}{\sum_i \cos\theta_{mean}(f_i) S_{\eta\eta}(f_i)}\right) \tag{8.56}$$

where, θ_m is a mean wave direction, and f_s is a sampling frequency.

In case that a mean wave direction cannot be obtained from the data, a mean wind direction may be considered as a good estimation for a mean wave direction. In other words, if a mean wave direction cannot be estimated directly, depending on the situation, a mean wave direction may be replaced by a mean wind direction.

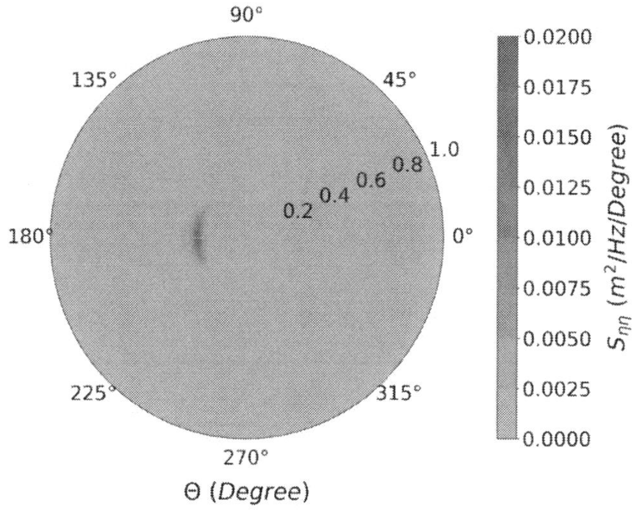

Figure 8.4 Schematic illustration of the water surface elevation directional power spectral density, $S_{\eta\eta}(f,\theta)$ in a polar format. A radial direction shows a frequency, and a colormap shows spectrum magnitudes. A directional spectrum is calculated from Donelan et al. (1985) for a wind direction of 180°. A sampling frequency is $f_s = 2\ Hz$. A schematic power spectral density is obtained from the JONSWAP spectrum (Hasselmann et al., 1973) for a peak wave frequency of 0.25 Hz.

Note that in the process of the velocity measurement, it is typical to collect data in the ENU (East-North-Up) coordinate system. When a wave direction is calculated for such cases, differences between the ENU coordinate system and a conventional Cartesian coordinate system (XYZ) should be considered.

Also, it needs to be pointed out that it is very common to report a wave direction following a meteorological convention. In a meteorological system, a direction is measured from the true North, increasing clockwise, i.e. a direction equal to 0°: means wave comes from the North, 90°: means wave comes from the East, 180°: means wave comes from the South, and 270°: means wave comes from the West.

There are other methods such as a Maximum Likelihood Method, MLM, and a Maximum Entropy Method, MEM, for a wave direction calculation (e.g. Benoit et al., 1997; Ewans, 1998; Massel and Brinkman, 1998; Earle et al., 1999; Donelan et al., 2015).

8.6 Generating a time series from a power spectral density

There are a series of empirical power spectral densities in the literature. Each of these spectrums was developed for certain conditions. For example, the JONSWAP spectrum was proposed for deep water (Hasselmann et al., 1973), and the TMA spectrum was proposed for intermediate and shallow water (Bouws et al., 1985). There are situations that it is required to generate a time series of data from a power spectral density. This time series construction can be from any of the empirical spectrums or from a spectrum obtained from measured data. In such cases, a time series can be generated by one of the following two methods.

In the first method, first, wave amplitudes associated with all frequencies in a spectrum are calculated. Then, a time series of the sinusoidal wave for each of these amplitudes and frequencies is generated. Finally, a time series associated with a spectrum is obtained by super-positioning all these sinusoidal wave time series. Steps to attain a time series from a spectrum are:

1. Define a frequency interval, df, in a power spectral density.
2. Define a sampling frequency, f_s, of the data. A power spectral density has a frequency range of $0 \leq f \leq f_s/2$, where f is a frequency.
3. Define a time interval, dt, as $dt = 1/f_s$.
4. Define the total number of data points, N. A power spectral density has $\frac{N}{2} + 1$ elements. A value of N should be a power of two.
5. Calculate a time vector (array), t, with a time range of $0 \leq t \leq (N-1)dt$ and a time interval of dt. A time vector has N elements.
6. Calculate wave height, $H(f)$, for all frequencies in a range of $0 \leq f \leq f_s/2$ as $H(f) = \left(4\sqrt{S_{\eta\eta}(f)df}\right)/\sqrt{2}$. It can be rewritten as $H(f) = 2\sqrt{2S_{\eta\eta}(f)df}$. Note that, $H \approx H_{rms} = H_{m0}/\sqrt{2}$, where H_{rms} is the root mean square (rms) wave height, $H_{m0} = 4\sqrt{m_0}$ is zero-moment wave height, $m_0 = \int_0^\infty S_{\eta\eta}(f)df$ is the zero-moment of the water surface elevation power spectral density, $S_{\eta\eta}$ is a water surface elevation power spectral density, and f is a frequency. Therefore, a wave

amplitude, a, for all frequencies in a range of $0 \le f \le f_s/2$ is calculated as $a(f) = H(f)/2 = \sqrt{2S_{\eta\eta}(f)df}$.

7. Generate a random phase, ϕ, between 0 and 2π. This randomly generated phase has $\frac{N}{2}+1$ elements associated with frequencies in a range of $0 \le f \le f_s/2$.
8. Calculate a wave angular frequency, ω, as $\omega = 2\pi f$ for frequencies in a range of $0 \le f \le f_s/2$.
9. Construct a time series for each wave amplitude as $\eta_n(t) = a_n \cos(-\omega_n t + \phi_n)$. The n is a counter which represents a value associated with each frequency in a frequency range of $0 \le f \le f_s/2$.
10. A water surface elevation time series, $\eta(t)$, is obtained from adding all calculated time series as $\eta(t) = \sum_{n=1}^{(N/2)+1} \eta_n(t)$. A time series of water surface elevation has N elements within a time range of $0 \le t \le (N-1)dt$, with a time interval of dt.

In the second method, first, waves amplitudes associated with all frequencies in a spectrum are calculated. Then, a two-sided Fourier transform is generated from these amplitudes. Finally, a time series associated with a spectrum is obtained by using an inverse Fourier transform. Steps to attain a time series from a spectrum are:

1. Define a frequency interval, df, in a power spectral density.
2. Define a sampling frequency, f_s, of the data. A power spectral density has a frequency range of $0 \le f \le f_s/2$, where f is a frequency.
3. Define a time interval, dt, as $dt = 1/f_s$.
4. Define the total number of data points, N. A power spectral density has $\frac{N}{2}+1$ elements. A value of N should be a power of two.
5. Convert a one-sided spectrum with a frequency range of $0 \le f \le f_s/2$ to two-sided spectrum a frequency range of $0 \le f \le (f_s/2) - df$. To do that, first, divide spectrum values with a frequency range of $0 < f < f_s/2$ by 2. Note that data points associated with $f = 0$ and $f = f_s/2$ are excluded and not divided by 2. Then, flip (mirror) the values of the spectrum with a frequency range of $0 < f < f_s/2$ against $f = f_s/2$. Again, note that data points associated with $f = 0$ and $f = f_s/2$ are excluded and are not flipped. A two-sided spectrum has N elements with a frequency range of $0 \le f \le (f_s/2) - df$.

6. Calculate an absolute value of Fourier transform, $r(f)$, for all frequencies in a range of $0 \le f \le (f_s/2) - df$ as $r(f) = N\sqrt{2S_{\eta\eta}(f)df}$.
7. Generate a random phase, ϕ, between 0 and 2π. This randomly generated phase has $\frac{N}{2} + 1$ elements associated with frequencies in a range of $0 \le f \le f_s/2$.
8. Flip (mirror) the values of the phase with a frequency range of $0 < f < f_s/2$ against $f = f_s/2$. Note that data points associated with $f = 0$ and $f = f_s/2$ are excluded and are not flipped. This constructs a symmetrical random phase which has N elements associated with frequencies in a range of $0 \le f \le (f_s/2) - df$.
9. Construct the Fourier transform signal as $\mathcal{F} = re^{i\phi}$ with a frequency range of $0 \le f \le (f_s/2) - df$. Note that $i = \sqrt{-1}$.
10. A water surface elevation time series, $\eta(t)$, is obtained from an inverse Fourier transform of $\mathcal{F}$. A time series of the water surface elevation has N elements within a time range of $0 \le t \le (N-1)dt$, with a time interval of dt.

The generated water surface elevation time series can be verified by calculating a power spectral density of this generated time series and comparing it with an original power spectral density. This calculated spectrum should be identical in terms of its properties such as m_0 with an original spectrum. A calculated spectrum might need to be smoothed to have the same shape as an original spectrum.

8.7 Wind Sea and swell waves partitioning

Under a normal condition, generally, two types of waves might exist in a water body, wind sea and swell waves. Wind sea refers to the waves that are generated locally, while swell waves refer to waves that are not generated locally and propagated from outside to an area of interest (e.g. Reeve et al., 2012). Depending on a condition, both of these waves or only one of them might be present in an area. There are areas where wind seas are the dominant waves while swell waves either are not existing or are negligible. On the contrary, there are areas where swell waves are dominant. In cases that both wind sea and swell waves with considerable energies are present in an area, a power spectral density of the measured water surface elevation contains and illustrates energies from both wind sea and swell waves. Wind sea generally has a smaller wavelength and a shorter wave period compared with swell waves. Because of that, in general, energies in a low-frequency section of the power spectral density are associated with swell waves, while energies in a high-frequency section of the spectrum are associated with wind sea.

In most studies, it is of general interest to separate wind sea energies from swell waves energies. A simple way to separate wind sea from swell waves is to consider the fact that a wave receives energy from wind if its phase speed, i.e. its celerity, is smaller than a wind velocity. In other words, waves that propagate faster than a wind velocity, do not receive energy from a local wind and most likely are swell waves. A wave phase speed or a wave celerity in deep water is equal to $C = gT/(2\pi) = g/(2\pi f)$, where C is a wave celerity, T is a wave period, f is a wave frequency, and g is the gravitational acceleration. Now, if a wave has a celerity larger than a wind velocity, i.e. if $C > U_w$ or $g/(2\pi f) > U_w$ where U_w is a wind velocity, then it is a swell wave, otherwise, it is wind sea. In other words, waves with frequencies that follow $f < g/(2\pi U_w)$ are swell waves and waves with frequencies that follow $f \geq g/(2\pi U_w)$ are wind sea. Therefore, a separation frequency, f_{sep}, which is a frequency that separates swell waves from wind sea is $f_{sep} = g/(2\pi U_w)$. All waves with frequencies $f < f_{sep}$ are considered swell waves and with frequencies $f > f_{sep}$ are considered wind waves.

However, more sophisticated methods are proposed in the literature to separate wind sea from swell waves. There are one-dimensional, $1D$, and two-dimensional, $2D$, methods. Also, there are wind based and non-wind based methods (e.g., Wang and Gilhousen, 1998; Gilhousen and Hervey, 2001; Portilla et al., 2009; Hwang et al., 2012). Gilhousen and Hervey (2001) proposed a $1D$ method to separate wind sea and swell waves as:

$$\xi(f) = \frac{H_{m0}(f)}{L(f)} = \frac{2\pi H_{m0}(f)}{gT_{m02}^2(f)} = \frac{8\pi m_2}{g\sqrt{m_0}} \tag{8.57}$$

$$f_{sep} = max\left(0.75\, f_x\, , 0.9\frac{1.25}{U_{10}}\right) \tag{8.58}$$

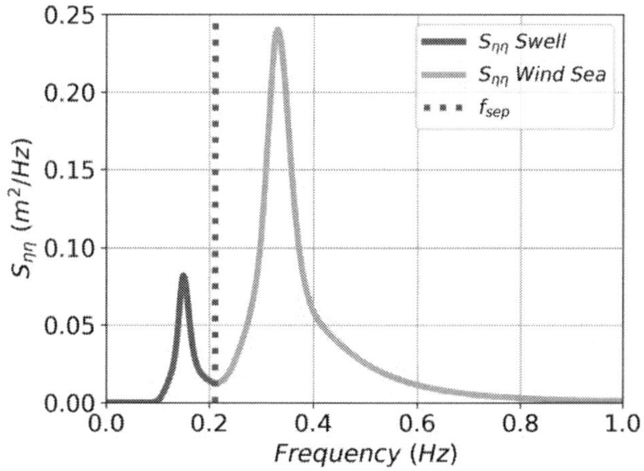

Figure 8.5 Schematic illustration of wind sea and swell waves separation using a separation frequency, f_{sep}. Plot shows a schematic water surface elevation power spectral density, $S_{\eta\eta}$, containing energies from both wind sea and swell waves. A separation frequency is $f_{sep} = 0.21\ Hz$ and a sampling frequency is $f_s = 2\ Hz$. Values of f_{sep} and f_s in the figure are selected for a demonstration purpose. In practice, these values need to be selected based on the physical properties of the study area, instrumentation, and nature of the research. A schematic power spectral density is obtained from a combination of the manipulated JONSWAP spectrum (Hasselmann et al., 1973) with peak wave frequencies of 0.15 and 0.33 Hz.

where, f_{sep} is a frequency that separates swell waves from wind sea, $\xi(f)$ is a wave steepness function, f is a frequency, f_x is a frequency associated with a maximum value of $\xi(f)$, L is a wavelength, U_{10} is a 10-minute averaged wind velocity at 10 m above a surface, $H_{m0} = 4\sqrt{m_0}$ is zero-moment wave height, $T_{m02} = \sqrt{m_0/m_2}$ is a mean wave period, m_0 is the zero-moment of the water surface elevation power spectral density, $m_n = \int_{f_l}^{f_u} f^n S_{\eta\eta}(f)\,df$ is the nth moment of the water surface elevation power spectral density within a frequency range of $f_l \le f \le f_u$, the f_l is a lower limit of the wave spectrum, the $f_u = 0.5\ Hz$ is an upper limit of the wave spectrum, and $S_{\eta\eta}$ is a water surface elevation power spectral density.

More recently, Hwang et al. (2012) proposed a method to separate wind sea and swell waves as:

$$I_1(f) = \frac{m_1}{\sqrt{m_{-1}}} \tag{8.59}$$

$$f_{sep} = 24.2084 f_{m1}^3 - 9.2021 f_{m1}^2 + 1.8906 f_{m1} - 0.04286 \tag{8.60}$$

Similarly, $m_n = \int_{f_l}^{f_u} f^n S_{\eta\eta}(f)\, df$ is the nth moment of the wave spectrum with $f_u = 0.5\ Hz$, and f_{m1} is a frequency associated with a maximum value of I_1.

Note that, in both Gilhousen and Hervey (2001) and Hwang et al. (2012) methods, a value of f_u might be required to be larger than $0.5\ Hz$ for waves in intermediate and shallow water.

9 Wave Modeling and Prediction

In this chapter, modeling and prediction of the coastal wave properties are described. First, wind and bathymetry data preparation are explained. Next, wave power spectra are discussed. Then, parametric wave models and parametric hurricane models are described. After that, a procedure to use SWAN numerical wave model is explained. Step by step procedure to generate SWAN input files and to run SWAN model are presented. Then, typical measures for the goodness of fit are discussed.

9.1 Non-dimensional variables

In oceanography and coastal science studies, it is a common practice to present parameters that describe wind, wave and physical domain in a non-dimension form. The most common non-dimensional variables, all denoted by the hat or caret symbol, ^, are (e.g., Sverdrup and Munk, 1947, Karimpour et al., 2017):

$$\hat{E} = \frac{g^2 m_0}{U_{10}^4} \tag{9.1}$$

$$\widehat{f_p} = \frac{f_p U_{10}}{g} \tag{9.2}$$

$$\hat{h} = \frac{gh}{U_{10}^2} \tag{9.3}$$

$$\hat{F} = \frac{gF}{U_{10}^2} \tag{9.4}$$

$$\hat{t} = \frac{gt}{U_{10}} \tag{9.5}$$

$$\widehat{k_p} = \frac{k_p U_{10}^2}{g} \tag{9.6}$$

where, $\hat{E}$ is dimensionless wave energy, $\widehat{f_p}$ is a dimensionless peak wave frequency, $\hat{h}$ is dimensionless water depth, $\hat{F}$ is a dimensionless wind fetch, $\hat{t}$ is dimensionless time, $\widehat{k_p}$ is a dimensionless peak wavenumber, m_0 is a zero-moment of the water surface elevation power spectral density, f_p is a peak wave frequency in Hz, h is water depth, F is a wind fetch, t is time in second, k_p is a peak wavenumber which is a wavenumber associated with a peak wave frequency, g is the gravitational acceleration, and U_{10} is a 10-minute averaged wind velocity at 10 m height from a surface. Although it is common to use U_{10} to make variables non-dimensional, any other wind velocity value can be used for that purpose as long as it stays consistent throughout the analysis.

In some studies, using mean water depth along a wind fetch is more suitable than using local water depth. In such cases, dimensionless water depth is:

$$\hat{h} = \frac{g\bar{h}}{U_{10}^2} \tag{9.7}$$

where, $\hat{h}$ is dimensionless water depth based on mean water depth. The $\bar{h}$ is mean water depth averaged along a wind fetch, which is calculated as:

$$\bar{h} = \frac{\int_0^F h(x)dx}{F} \approx \frac{\sum_i h_i \Delta x_i}{\sum_i \Delta x_i} \tag{9.8}$$

where, $h(x)$ is spatial water depth along a wind fetch axis, and x is a longitudinal axis along a wind fetch. If a water body has a horizontal bathymetry, i.e. horizontal bottom, then $\bar{h} = h$ where h is local water depth at a study location.

9.2 Geographic coordinate system

In most wave models, physical properties of the water body, such as a domain boundary, are given as a function of the earth coordinate system. Conventionally, a geographic coordinate system is presented by latitude, ϕ, and longitude, λ.

A latitude, ϕ, shows an angle that a line originated in earth center and passes through a point on a surface of the earth has with an equatorial plane. It is $0°$ at the equator and is ranged from $90\ S$ or $-90°$ at the south pole to $90\ N$ or $90°$ at the north pole.

A longitude, λ, shows an angle that a meridian of a point on a surface of the earth has with a reference meridian. It is $0°$ at a reference meridian and is ranged from $180\ W$ or $-180°$ to $180\ E$ or $180°$.

A great circle distance (or an arc distance) between points on the earth surface is required in many studies, such as in the real-world wave data analysis and modeling. A great circle distance shows the shortest distance between two points on a surface of the sphere. A great circle distance between two points with coordinates of (ϕ_1, λ_1) and (ϕ_2, λ_2) is calculated as:

$$dist = R \times \Delta\sigma \tag{9.9}$$

where, $dist$ is a distance between two points, which represents a length of the arc that passes through the points, and R is the earth radius. A mean radius of the earth is $R = 6.371 \times 10^6\ m$. The $\Delta\sigma$ is a central angle between points.

There are different methods to calculate a central angle. One way to obtain a central angle, $\Delta\sigma$, is to use the spherical law of cosines as:

$$\Delta\sigma = \arccos(\sin\phi_1 \sin\phi_2 + \cos\phi_1 \cos\phi_2 \cos(\Delta\lambda)) \tag{9.10}$$

where, $\Delta\lambda = \lambda_2 - \lambda_1$.

Another method to find a central angle, $\Delta\sigma$, is to use the Haversine formula as:

$$A = \sin^2\left(\frac{\Delta\phi}{2}\right) + \cos\phi_1 \cos\phi_2 \sin^2\left(\frac{\Delta\lambda}{2}\right) \tag{9.11}$$

$$\Delta\sigma = 2\,\text{arctan2}\left(\sqrt{A}, \sqrt{1-A}\right) \tag{9.12}$$

where, $\Delta\phi = \phi_2 - \phi_1$. Note that, the Haversine formula suffers from an inaccuracy for antipodal points.

Another method to calculate a central angle, $\Delta\sigma$, is to use the Vincenty formula (Vincenty, 1975) as:

$$B = \sqrt{(\cos\phi_2 \sin(\Delta\lambda))^2 + (\cos\phi_1 \sin\phi_2 + \sin\phi_1 \cos\phi_2 \cos(\Delta\lambda))^2} \tag{9.13}$$

$$C = \sin\phi_1 \sin\phi_2 + \cos\phi_1 \cos\phi_2 \cos(\Delta\lambda) \tag{9.14}$$

$$\Delta\sigma = \text{arctan2}(B, C) \tag{9.15}$$

where, $\Delta\lambda = \lambda_2 - \lambda_1$. Note that, the Vincenty formula is accurate in all locations.

Similar to a great circle distance, azimuth (or a bearing) of the points on the earth surface is usually required in different studies. Azimuth shows an angle measured clockwise from the true North, i.e. azimuth equal to 0°: means toward the North, 90°: means toward the East, 180°: means toward the South, and 270°: means toward the West. Azimuth should not be mistaken for a conventional meteorological direction, which a direction equal to 0°: means from the North, 90°: means from the East, 180°: means from the South, and 270°: means from the West. Azimuth can be converted to a conventional meteorological direction by adding 180° to azimuth value. After a converted value is calculated, add 360° or π to any negative value to make it positive, and subtract 360° or π from any value that is larger than 360° or π. This keeps all values between 0° and 360° or between 0 and π. Azimuth or a bearing between two points with coordinates of (ϕ_1, λ_1) and (ϕ_2, λ_2) is calculated as:

$$Az = \text{arctan2}(\sin(\Delta\lambda)\cos\phi_2\,, \cos\phi_1 \sin\phi_2 - \sin\phi_1 \cos\phi_2 \cos(\Delta\lambda)) \tag{9.16}$$

where, *Az* is azimuth or a bearing which is an angle measured clockwise from the true North.

There are cases that a start-point of the arc, an arc length, and azimuth of the arc are all known, however, an endpoint of that arc is not known. In such cases, coordinates of the arc endpoint, i.e. (ϕ_2, λ_2), can be found from a start point of the arc, (ϕ_1, λ_1), an arc length, $dist$, and azimuth, Az, as:

$$\phi_2 = \arcsin(\sin\phi_1 \cos(\Delta\sigma) + \cos\phi_1 \sin(\Delta\sigma)\cos(Az)) \tag{9.17}$$

$$\lambda_2 = \lambda_1 + \text{arctan2}(\sin(Az)\sin(\Delta\sigma)\cos\phi_1\, , \cos(\Delta\sigma) - \sin\phi_1 \sin\phi_2) \tag{9.18}$$

where, (ϕ_1, λ_1) are coordinates of the arc start-point, $dist$ is an arc length, Az is azimuth, and $\Delta\sigma = dist/R$.

It should be pointed out that in any numerical calculations, values of $\phi_1, \lambda_1, \phi_2, \lambda_2, \Delta\sigma$, and Az should to be in radian.

9.3 Wind data

Wind data are one of the main parameters required for wave modeling, simulation, hindcasting, and prediction. Wind data often contain two sets of values, one for wind velocity magnitudes and one for wind directions. Typically, wind data are measured and reported at specific intervals, such as every hour or every half an hour. For example, if data are measured and reported hourly, then a wind dataset contains one value for wind velocity and one value for a wind direction at the beginning of each hour.

It needs to be pointed out that it is very common to report a wind direction following a meteorological convention. In a meteorological system, a direction is measured clockwise from the true North, i.e. a direction equal to 0°: means wind comes from the North, 90°: means wind comes from the East, 180°: means wind comes from the South, and 270°: means wind comes from the West. To convert a meteorological direction to a trigonometric angle, multiply a meteorological direction by -1 and then add 270° to it. After a converted value is calculated, add $360°$ or π to any negative value to make it positive, and subtract $360°$ or π from any value that is larger than $360°$ or π. This keeps all values between $0°$ and $360°$ or between 0 and π.

Generally, wind data are obtained from weather stations. Each of these weather stations may collect wind data with different settings than other stations. Because of that, wind datasets from different stations may not be similar. Therefore, to make wind data from different stations comparable, measured wind data need to be standardized. This standardization helps that all researchers use comparable wind data (e.g. Masters et al., 2010; He et al., 2014).

Among measurement settings and conditions, two factors are more important than others. One is a wind averaging duration and the other is a wind measurement height. The most acceptable standard for a wind averaging duration is 10 minutes, and for a wind measurement height is 10 m from a surface. Besides these two, averaging a wind direction requires particular attention to avoid an error in a mean direction calculation around 0°. In following sections, it is explained how to convert wind data to a standard condition if a wind measurement does not meet standard conditions.

9.3.1 Wind gust factor (wind velocity averaging)

It was mentioned that wind data are measured and reported at a specific interval such as every hour. To get a data point at a beginning of each interval, wind velocity and wind direction are measured for a specific period of time, and then averaged over that measurement period. For example, if wind data are reported every hour, wind velocity and wind direction at a beginning of every hour are obtained by measuring and averaging wind data for a specific duration, say 10-minutes, at a beginning that hour. It is a common practice to measure wind data for two or ten minutes at a beginning of each reporting interval and then average them over that period. Duration of this measurement at a beginning of each interval affects the velocity outcomes. Generally, as a measurement duration increases, averaged wind velocity decreases. For example, for the same wind, average velocity obtained from the 2-minutes long measurement is usually larger than average velocity obtained from the 10-minutes long measurement.

The most acceptable standard for the averaging duration of the wind is 10 minutes. Therefore, a majority of the wind theories and models are developed

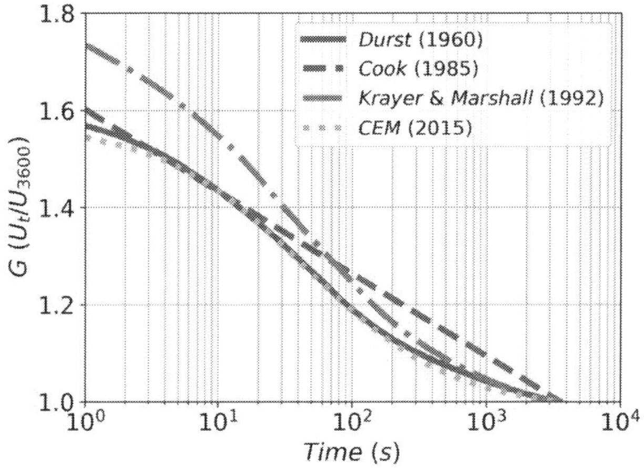

Figure 9.1 Comparison of wind gust factors, G, from Durst (1960), Cook (1985), Krayer and Marshall (1992), and Coastal Engineering Manual, CEM, (2015). The Durst curve is based on ASCE-7 (2005). The Cook curve is presented for a wind longitudinal turbulence intensity of $I_u = 0.175$. All curves are presented for $0 \leq t \leq 3600\ s$.

based on wind velocity averaged over the 10-minutes period. In other words, 10-minute averaged wind velocity is used in most wave models. Because of that, if wind velocities are measured and averaged over a different duration than 10 minutes, then wind velocities need to be converted to 10-minute averaged wind velocities. There are different methods proposed in the literature for converting averaged wind velocity obtained from a specific duration to any given duration (e.g. Durst, 1960; Deacon, 1965; Wieringa, 1973; Greenway, 1979; Cook, 1985; Mitsuta and Tsukamoto, 1989; Krayer and Marshall, 1992; Ashcroft, 1994; Verkaik, 2000; ESDU, 2002; Vickery and Skerlj, 2005; Harper et al., 2010; Masters et al., 2010; He et al., 2014; Holmes et al., 2014; Shu et al., 2015; Bardal and Sætran, 2016).

In the wind engineering literature, it is typical to show a wind velocity ratio respect to 1-hour averaged wind velocity as:

$$G(t, 3600) = \frac{U_t}{U_{3600}} \tag{9.19}$$

where, G is a wind gust factor, U_{3600} is wind velocity averaged over a duration of 3600 seconds (1 hour), U_t is wind velocity averaged over a duration of t seconds, the t is time and shows the desired wind velocity averaging duration in seconds. By knowing a wind gust factor, wind velocity can be converted from one averaging duration to another averaging duration.

A wind gust factor which shows gust wind velocity respect to sustained mean wind velocity often presented as (e.g. Mitsuta and Tsukamoto, 1989):

$$G = 1 + g_t \frac{\sigma_u}{\bar{U}} \tag{9.20}$$

where, g_t is a peak factor, σ_u is a standard deviation of the longitudinal wind velocity, and $\bar{U}$ is mean wind velocity.

There are several theoretical and empirical relationships in the literature that aimed for calculating a wind gust factor. One of the earliest relationships is proposed by Durst (1960), which is used as the Durst curve by the American Society of Civil Engineers, ASCE (ASCE-7, 2005). The Durst curve in ASCE-7 (2005) can be approximated by:

$$G(t, 3600) = 0.977 + \frac{0.64}{1+\left(\frac{t}{38.14}\right)^{0.685}} \tag{9.21}$$

Cook (1985) proposed an approximated relationship for a wind gust factor based on Wieringa (1973) as:

$$G(t, 3600) = 1 + 0.42 I_u \ln\left(\frac{3600}{t_w}\right) \tag{9.22}$$

where, $I_u = \sigma_u / \bar{U}$ is a wind longitudinal turbulence intensity. There are relationships in the wind engineering literature that calculate I_u based on the height and surface roughness, such as $I_u = \frac{1}{\ln(z/z_0)}$, where, z is height, and z_0 is surface roughness (e.g. Harper et al., 2010; Kwon and Kareem, 2013). Ishizaki (1983) and Cao et al. (2009) suggested similar relationships with different coefficients.

Krayer and Marshall (1992) proposed a curve for calculating a wind gust factor in a hurricane. The Krayer and Marshall curve can be approximated by:

$$G(t,3600) = 0.96 + \frac{0.839}{1+\left(\frac{t}{36.27}\right)^{0.655}} \tag{9.23}$$

Coastal Engineering Manual, CEM, (2015) suggests following relationship for converting wind velocity from one averaging duration to another averaging duration:

$$G(t,3600) = 1.277 + 0.296\tanh\left(0.9\log\left(\frac{45}{t}\right)\right) \quad \text{for } 1\,s \le t \le 3600\,s \tag{9.24}$$

$$G(t,3600) = 1.5334 - 0.15\log(t) \qquad \text{for} \quad 3600\,s < t < 36000\,s \tag{9.25}$$

If the measured wind is averaged over a one-hour period, then any of these relationships can be used to convert wind velocity to the intended duration. However, if the measured wind is not averaged over a one-hour period, first, wind velocity needs to be converted to U_{3600}, and then, U_{3600} is converted to the intended averaging duration. This can be done by calculating a wind gust factor for both durations and then use the ratio as a conversion factor. For example, to convert wind velocity averaged over a 2-minutes period, i.e. U_{120}, to wind velocity averaged over a 10-minutes period, i.e. U_{600}, by using a CEM (2015) gust factor, first, a gust factor for $t = 120\,s$ and $t = 600\,s$ are calculated as $G(120{,}3600) \approx 1.168$ and $G(600{,}3600) \approx 1.05$, respectively. Then, a conversion factor from $t = 120\,s$ to $t = 600\,s$ is $G(600{,}120) = U_{600}/U_{120} = G(600{,}3600)/G(120{,}3600) \approx 1.05/1.168 \approx 0.898$.

Note that, it is assumed that an averaging duration does not have any effect on a wind direction. Therefore, a wind gust factor is only used for wind velocity magnitude.

9.3.2 Wind velocity profile

Another factor that affects wind velocity is a measurement height. Generally, as a measurement height increases, wind velocity increases as well. In other words, wind velocity increases with height. The most acceptable standard for a wind height measurement is 10 m above a surface. Therefore, a majority of the wind theories and models are developed based on wind velocity measured at 10 m height. In other words, wind velocity at 10 m height is used in most wave models. Because of that, if wind velocity is measured at a different height than 10 m, then it needs to be converted to a velocity at 10 m height from a surface. To do that, a theoretical wind velocity profile is used.

Typically, there are two widely-used velocity profiles, a power law velocity profile and a logarithmic velocity profile (e.g. Holmes, 2015). Although a logarithmic velocity profile is more accurate, a power law velocity profile, due to its simplicity, is mostly used to convert wind velocity from one height to another height. An empirical power law velocity profile is written as:

$$\frac{U(z_1)}{U(z_2)} = \left(\frac{z_1}{z_2}\right)^{\alpha} \tag{9.26}$$

where, $U(z_1)$ and $U(z_2)$ are wind velocities at the height of z_1 and z_1 from a surface, respectively. In the wind engineering literature, a power law wind velocity profile is often written respect to the wind at 10 m height as:

$$\frac{U(z)}{U_{10}} = \left(\frac{z}{10}\right)^{\alpha} \tag{9.27}$$

where, $U(z)$ is wind velocity at the height of z from a surface, the U_{10} is wind velocity measured at 10 m height above a surface, and z is an upward vertical axis with zero at a surface. Here, z shows a height from a surface. The α is a power law exponent.

A value of α depends on terrain roughness. There is a range of recommendations for α in the literature (e.g. ASHRAE 2009; Tamura and Kareem, 2013). For example, typical values suggested for α are $\alpha = 1/11.5$ for water, $\alpha = 1/9.5$ for an open area, $\alpha = 1/7$ for a suburban area, and $\alpha = 1/5$ for an urban area (e.g. Davenport, 1960; ASCE-7, 2005). For a practical purpose, α is suggested as $\alpha = 1/7$.

A logarithmic velocity profile, which is more accurate than a power law velocity profile, is written as:

$$\frac{U(z)}{u_*} = \frac{1}{\kappa} \ln\left(\frac{z - z_d}{z_0}\right) \tag{9.28}$$

where, $U(z)$ is wind velocity at the height of z from a surface, z is an upward vertical axis with zero at a surface, $u_* = \sqrt{\tau/\rho}$ is friction velocity (a so-called shear velocity), τ is a surface shear stress, ρ is a fluid density (here, ρ is air density), $\kappa \approx 0.41$ is the von Karman's constant, z_0 is surface roughness, and z_d is a zero-plane displacement distance.

In the flat ground, wind velocity becomes zero at a ground surface, however, in dense terrain, a wind velocity profile shifts upward. It means that, in dense terrain, wind velocity becomes zero at the height of z_d instead of becoming zero at a ground surface. The z_d shows an amount that a zero-velocity height shifts upward in dense terrain. Similar to α in a power law profile, there is a range of recommendations for z_0, u_* and z_d for different terrains in the literature (e.g. SACE-7, 2005; Simiu, 2011; Emeis 2012; Tamura and Kareem, 2013). For example, typical values suggested for z_0 are $z_0 = 0.005\,m$ for water, $z_0 = 0.02\,m$ for an open area, $z_0 = 0.3\,m$ for a suburban area, and $z_0 = 2\,m$ for an urban area (e.g. ASCE-7, 2005; Simiu, 2011). Karimpour et al. (2012) described the details on how to calculate z_0, u_* and z_d from wind velocity profile data.

For wind over an ocean, a zero-plane displacement distance is assumed to be zero, i.e. $z_d = 0$. Charnock (1955) proposed surface roughness over an ocean (water) as:

$$z_0 = \alpha_{Ch} \frac{u_*^2}{g} \tag{9.29}$$

where, α_{Ch} is an empirical constant (a so-called the Charnock parameter). The α_{Ch} generally has a range between 0.01 and 0.02, although higher values are reported (e.g. Fairall et al., 1996; Letchford and Zachry, 2009; Holmes, 2015; Fisher et al., 2015). Charnock (1955) suggested $\alpha_{Ch} = 0.012$ and Wu (1980) suggested $\alpha_{Ch} = 0.0185$. However, the most used value for α_{Ch} was suggested by Garratt (1977) as $\alpha_{Ch} = 0.0145$.

A power law exponent, α, can be related to surface roughness, z_0, as (e.g. Emeis, 2012; Holmes, 2015):

$$\alpha = \frac{1}{\ln(z_{ref}/z_0)} \tag{9.30}$$

where, z_{ref} is a reference height. The z_{ref} shows a height that a power law profile and a logarithmic profile are matched at that height.

9.3.3 Wind direction averaging

Unlike wind velocity, it is assumed that an averaging duration and a measurement height do not have any effects on a wind direction. However, a simple averaging method cannot be used to average wind direction values. To explain the reason, assume there are two wind direction readings, one equal to 5° and the other equal to 355°. A correct average direction for these two values is 0°. However, using a simple average for these two values leads to $(5° + 355°)/2 = 180°$, which is obviously wrong. This occurs because a wind direction is a circular function as 0° direction is the same as 360° direction (2π direction). Therefore, a specific procedure needs to be followed to avoid inaccuracy in averaged wind directions, particularly the ones fluctuate around 0° or 360° (e.g. Mori, 1986; Bailey, 2000; Farrugia and Micallef, 2006). Following a method proposed by Yamartino (1984), wind directions are averaged as:

$$\bar{\theta} = \tan^{-1}\left(\frac{(\sum_{i=1}^{N} \sin\theta_i)/N}{(\sum_{i=1}^{N} \cos\theta_i)/N}\right) \tag{9.31}$$

where, θ is a direction, $\bar{\theta}$ is a mean direction, N is the total number of data points. To use this equation in a programming language, it is common to use the arctan2 function instead of the regular arctangent function (i.e. instead of the regular inverse tangent function). After $\bar{\theta}$ is calculated, add 360° or π to any negative value to make it positive, and subtract 360° or π from any value that is larger than 360° or π. This keeps all values between 0° and 360° or between 0 and π.

9.4 Sustained wind

One of the wind criteria that is required in some wave models, such as parametric wave models, is the sustainability of the wind. The wind is

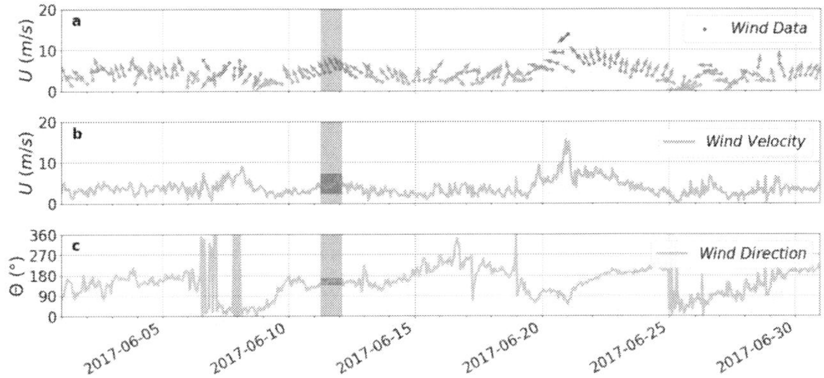

Figure 9.2 Sustained wind duration demonstration. Plot (a) shows wind velocities in a vector form. Plot (b) shows wind velocity values. Plot (c) shows wind velocity directions. A sample of sustained wind is shown by a vertical window in all three plots. In these windows, wind velocity varies less than 2.5 m/s, and a wind direction varies less than 15°, following a requirement by Coastal Engineering Manual, CEM, (2015). Wind data are obtained from NOAA's Tides and Currents (tidesandcurrents.noaa.gov), at Grand Isle station (Station ID: 8761724) from Jun 1, 2017 to Jun 30, 2017 UTC, in 1 hr interval.

considered to be sustained and steady if a wind magnitude and a wind direction, both, are steady (e.g. Karimpour et al., 2017). Coastal Engineering Manual, CEM, (2015) considers wind to be steady if both of the following relationships are valid at the same time:

$$|U_i - \overline{U}| \leq 2.5\ m/s \tag{9.32}$$

and

$$\left|\theta_i - \bar{\theta}\right| \leq 15° \tag{9.33}$$

where, U_i is the ith data point in a wind velocity dataset, $\overline{U}$ is mean wind velocity, θ_i is the ith data point in a wind direction dataset, and $\bar{\theta}$ is a mean wind direction. To obtain mean values of wind velocity, $\overline{U}$, and wind direction, $\bar{\theta}$, values of velocities and directions are averaged over all of the preceding consecutive data points that consecutively satisfied both of these sustainability conditions (e.g. Karimpour et al., 2017).

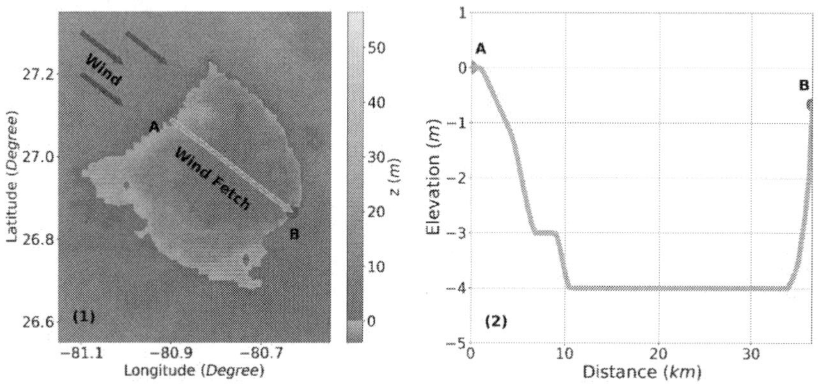

Figure 9.3 A wind fetch schematic illustration. A wind fetch over a lake is illustrated for the point B as the wind comes from the North-West. A wind fetch starts from an outside boundary, point A, toward a point of interest, point B. A position of the point A on a faraway boundary is imposed by a wind direction in such a way that the line AB maintains the same direction as a wind direction. A wind fetch at point B for wind coming from the North-West is equal to a length of the line AB. If point B is located at any other location, point A and line AB are similarly defined based on a wind direction. Plot (2) shows a bathymetry profile along a wind fetch from point A to point B. Data in this figure shows Lake Okeechobee in Florida, USA. Elevation data are extracted from ETOPO1 Global Relief Model (Amante and Eakins, 2009).

9.5 Wind fetch

A wind fetch is one of the parameters that is required in some wave models, such as parametric wave models. A wind fetch shows a distance that wind has traveled over open water without facing any obstacle, such as land, until reaching a point of interest. Typically, a wind fetch is measured up to a shoreline. However, if a point of interest is in the middle of the water body, then it is measured up to that point.

A wind fetch is generally measured along a wind direction. It is based on an assumption that a wave is generated and travels along a wind direction (i.e. assuming a wave direction is the same as a wind direction). To measure a wind fetch, a line is drawn with the same direction as a wind direction from

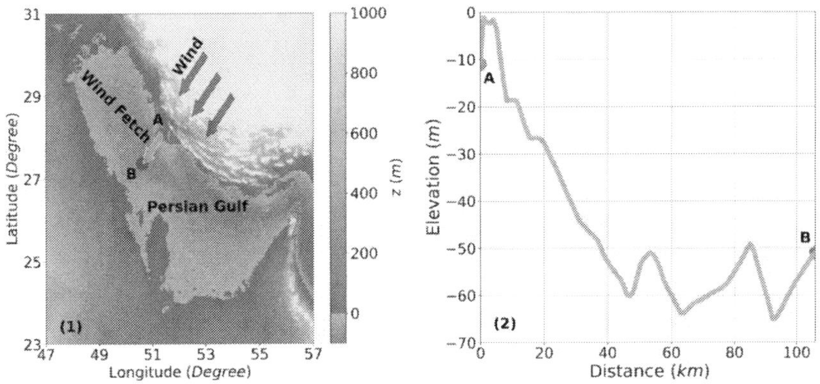

Figure 9.4 A wind fetch schematic illustration. A wind fetch over the Persian Gulf is illustrated for the point B as the wind comes from the North-East. A wind fetch starts from an outside boundary, point A, toward a point of interest, point B. A position of the point A on a faraway boundary is imposed by a wind direction in such a way that the line AB maintains the same direction as a wind direction. A wind fetch at point B for wind coming from the North-East is equal to a length of the line AB. If point B is located at any other location, point A and line AB are similarly defined based on a wind direction. Plot (2) shows a bathymetry profile along a wind fetch from point A to point B. Data in this figure shows the Persian Gulf. Elevations in figure are limited from -100 m to 1000 m. Elevation data are extracted from ETOPO1 Global Relief Model (Amante and Eakins, 2009).

an outside boundary, such as a faraway shoreline, toward the point of interest. A Length of this line is a wind fetch.

If a wind fetch is used for a wave calculation, it is important that the wind has enough sustained duration to let a wave travels a full length of the wind fetch. In other words, a duration of the sustained wind in a direction of the wind fetch should be long enough to provide sufficient time for wind waves to travel a full fetch and reach a destination. In this case, wind waves are called fetch-limited. However, if a sustained wind duration is not long enough to allow wind waves travel a full fetch, then waves are duration-limited (e.g. Karimpour et al., 2017). In case that wave growth is duration-limited, an equivalent fetch may be required to be calculated for a duration of sustained wind.

In cases that a sustained wind duration is long enough and wind growth is fetch-limited, then full fetch values are used in wave calculations. Otherwise, if a sustained wind duration is not long enough to let a wind wave travels a full fetch and wave growth becomes duration-limited, then an equivalent fetch instead of a full fetch, based on situations, might need to be used in wave calculations.

9.6 Bathymetry data

Bathymetry data are one of the main inputs required for wave modeling, simulation, hindcasting, and prediction. Bathymetry data are available in different formats such as a geo-referenced tiff (GeoTIFF), netCDF, and XYZ gridded ASCII format. Bathymetry data mostly contains three sets of values, one for longitudes, one for latitudes, and one for elevations associated with these latitudes and longitudes.

There are situations that bathymetry data cannot be used in their original format and need to be translated to a new format. Bathymetry data can be read and translated to other formats either by using Geographic Information System, GIS, software such as ArcGIS (Esri, www.esri.com), QGIS (Open Source Geospatial Foundation, www.qgis.org), and GRASS GIS (Open Source Geospatial Foundation, grass.osgeo.org), or by using a mapping and GIS toolbox in programming languages such MATLAB (The MathWorks, Inc., www.mathworks.com) and Python (Python Software Foundation; www.python.org). One of the most popular libraries that can be used to handle bathymetry data is Geospatial Data Abstraction Library (GDAL, Open Source Geospatial Foundation, www.gdal.org), which has a wide range of tools for reading, writing, and manipulating geospatial data.

Some of the well-known digital elevation models, DEM, and sources for geospatial, topography, bathymetry, relief, and lidar datasets are listed in the following tables.

Table 9.1 Digital elevation model (DEM)

Digital Elevation Model Name	Digital Elevation Model Address
NOAA global relief model (ETOPO1)	www.ngdc.noaa.gov/mgg/global/global.html
USGS Global Multi-resolution Terrain Elevation (GMTED2010)	lta.cr.usgs.gov/GMTED2010 topotools.cr.usgs.gov/gmted_viewer
USGS Global 30 Arc-Second Elevation (GTOPO30)	lta.cr.usgs.gov/GTOPO30
NASA Advanced Spaceborne Thermal Emission and Reflection Radiometer (ASTER)	asterweb.jpl.nasa.gov
NASA Shuttle Radar Topography Mission (SRTM)	www2.jpl.nasa.gov/srtm
GEBCO Gridded bathymetry data	www.gebco.net
ALOS Global Digital Surface Model (AW3D30)	www.eorc.jaxa.jp/ALOS/en/aw3d30

USGS: The United States Geological Survey
NOAA: National Oceanic and Atmospheric Administration
NASA: National Aeronautics and Space Administration
GEBCO: General Bathymetric Chart of the Oceans
ALOS: Advanced Land Observing Satellite

9.7 Water wave power spectra

It is general interest of researchers and engineers to estimate wave properties in a given water body, such as an ocean and a sea. One method to estimate wave properties in a water body is to present wave properties with a water surface elevation power spectral density, $S_{\eta\eta}$, for a random sea. One dimensional wave spectrum can be expressed as (e.g. Phillips, 1958; Ochi, 2005; Sorensen, 2005; Holthuijsen, 2007; McCormick, 2009; Reeve et al., 2012):

$$S_{\eta\eta}(f) = \frac{A}{f^5}\exp\left(-\frac{B}{f^4}\right) \tag{9.34}$$

where, $S_{\eta\eta}$ is a water surface elevation power spectral density, f is a frequency in Hz, and A and B are spectrum shape and scale coefficients. For a given wave spectrum, wave properties are estimated as (e.g. McCormick, 2009; Reeve et al., 2012):

$$m_n = \int_0^\infty S_{\eta\eta}(f) f^n df = \frac{A}{4B^{\left(1-\frac{n}{4}\right)}}\Gamma\left(1-\frac{n}{4}\right) \tag{9.35}$$

$$m_0 = \int_0^\infty S_{\eta\eta}(f) df = \frac{A}{4B} \tag{9.36}$$

$$m_1 = \int_0^\infty S_{\eta\eta}(f) f df = 1.2254\frac{A}{4B^{\left(\frac{3}{4}\right)}} \tag{9.37}$$

Table 9.2 Geospatial, topography, bathymetry, relief, and lidar dataset

Data Source Name	Data Source Address
Natural Earth	www.naturalearthdata.com
USGS Earth Explorer	earthexplorer.usgs.gov
USGS Global Multi-resolution Terrain Elevation (GMTED2010)	lta.cr.usgs.gov/GMTED2010 topotools.cr.usgs.gov/gmted_viewer
USGS Global 30 Arc-Second Elevation (GTOPO30)	lta.cr.usgs.gov/GTOPO30
USGS Coastal National Elevation Database (CoNED)	topotools.cr.usgs.gov/coned
Elevation Derivatives for National Applications (EDNA)	edna.usgs.gov lta.cr.usgs.gov/edna
NOAA ETOPO1 global relief model	www.ngdc.noaa.gov/mgg/global/global.html www.ngdc.noaa.gov/mgg/global/relief/ETOPO1/data
NOAA grid extract tool	maps.ngdc.noaa.gov/viewers/wcs-client
NOAA U.S. Coastal Relief Model	www.ngdc.noaa.gov/mgg/coastal/crm.html
NOAA VDatum DEM Project	www.ngdc.noaa.gov/mgg/inundation/vdatum/vdatum.html
The National Map	nationalmap.gov
GEBCO	www.gebco.net
British Oceanographic Data Centre	https://www.bodc.ac.uk
NOAA Data Access Viewer	coast.noaa.gov/dataviewer
NOAA Interagency Elevation Inventory	coast.noaa.gov/inventory
OpenTopography	www.opentopography.org
OpenLayers	openlayers.org

USGS: The United States Geological Survey
NOAA: National Oceanic and Atmospheric Administration
DEM: Digital Elevation Model
GEBCO: General Bathymetric Chart of the Oceans

$$m_2 = \int_0^\infty S_{\eta\eta}(f) f^2 df = 1.77245 \frac{A}{4B^{\left(\frac{1}{2}\right)}} \quad (9.38)$$

$$H_{m0} = 4\sqrt{m_0} = 4\sqrt{\int_0^\infty S_{\eta\eta}(f) df} = 4\sqrt{\frac{A}{4B}} \quad (9.39)$$

$$T_p = \frac{1}{f_p} \approx 1.057 B^{(-1/4)} \quad (9.40)$$

$$T_{m01} = \frac{m_0}{m_1} \approx 0.816 B^{(-1/4)} \quad (9.41)$$

$$T_{m02} = \sqrt{\frac{m_0}{m_2}} \approx 0.751 B^{(-1/4)} \quad (9.42)$$

where, m_n is the nth moment of the wave power spectrum, m_0, m_{01} and m_{02} are the zero, first and second moment of the wave power spectrum, respectively, H_{m0} is zero-moment wave height, T_p is a peak wave period, T_{m01} is a mean wave period, T_{m02} is a mean zero-crossing period, and f_p is a peak wave frequency in Hz, which is a frequency associated with a maximum value of the $S_{\eta\eta}$. In practice, a peak wave frequency, f_p, is obtained by locating a frequency associated with the maximum value of the $S_{\eta\eta}$.

There is a rich literature on wave power spectra in deep and shallow water. Here, some of the most prominent water surface elevation power spectral densities, $S_{\eta\eta}$, in both deep and shallow water are presented.

9.7.1 Bretschneider (1959) spectrum in deep water

One of the earliest water surface elevation power spectral densities in deep water is suggested by Bretschneider (1959, 1965) as:

$$S_{\eta\eta}(T) = 3.44\frac{\bar{H}^2}{\bar{T}^4}T^3 \exp\left(-0.675\left(\frac{T}{\bar{T}}\right)^4\right) \tag{9.43}$$

where, T is a wave period, $\bar{T}$ is a mean wave period, and $\bar{H}$ is mean wave height. Bretschneider (1977) proposed that $T_s = (4/5)^{(1/4)}T_p$, where, T_s is a significant wave period, and T_p is a peak wave period.

9.7.2 Pierson and Moskowitz (1964) spectrum in fully developed deep water

Pierson and Moskowitz (1964) proposed a water surface elevation power spectral density for fully developed deep-water waves based on the wind and waves data from British weather ships in the North Atlantic as:

$$S_{\eta\eta}(f) = \frac{\alpha g^2}{(2\pi)^4 f^5}\exp\left(-1.25\left(\frac{f_p}{f}\right)^4\right) \tag{9.44}$$

where, $\alpha = 8.1 \times 10^{-3}$, the f is a frequency in Hz, the f_p is a peak wave frequency in Hz, and g is the gravitational acceleration. Based on the Pierson and Moskowitz spectrum, Ochi (1982) suggested that a peak wave frequency, f_p, and zero-moment wave height, H_{m0}, are:

$$f_p = \frac{0.21U_{19.5}^2}{g} \quad (9.45)$$

$$H_{m0} = \frac{0.87g}{2\pi U_{19.5}} \quad (9.46)$$

where, $U_{19.5}$ is wind velocity measured at 19.5 m above a surface. As mentioned above, wind velocity is mostly measured at 10 m above a surface. In that case, $U_{19.5}$ can be obtained from a conversion of the wind velocity at 10 m above a surface, U_{10}, to $U_{19.5}$. A simple method to do that is to use a power law velocity profile as $U(z)/U_{10} = (z/10)^{\alpha}$, where, $U(z)$ is wind velocity at height of z, and α is a power law exponent with typical values of $\alpha = 1/11.5$ for water, $\alpha = 1/9.5$ for an open area, $\alpha = 1/7$ for a suburban area, and $\alpha = 1/5$ for an urban area (e.g. Davenport, 1960; ASCE-7, 2005). By using this relationship, $U_{19.5}$ is calculated from U_{10} as $U_{19.5} = 1.95^{\alpha} U_{10}$.

Note that the Pierson and Moskowitz spectrum is developed and only suggested for a fully developed sea (wave field) in deep water.

9.7.3 JONSWAP (1973) spectrum in deep water

One of the most well-known and widely used water surface elevation power spectral densities in deep water is the JONSWAP spectrum. The JONSWAP spectrum was developed during the Joint North Sea Wave Project (JONSWAP) by Hasselmann et al. (1973) as:

$$S_{\eta\eta}(f) = \frac{\alpha g^2}{(2\pi)^4 f^5} \exp\left(-1.25\left(\frac{f_p}{f}\right)^4\right) \gamma^{\exp\left(-\frac{(f-f_p)^2}{2\sigma^2 f_p^2}\right)} \quad (9.47)$$

$$\alpha = 0.076\left(\frac{gF}{U_{10}^2}\right)^{-0.22} = 0.076\hat{F}^{-0.22} \quad (9.48)$$

$$f_p = \frac{3.5g}{U_{10}}\left(\frac{gF}{U_{10}^2}\right)^{-0.33} = \frac{3.5g}{U_{10}}\hat{F}^{-0.33} \quad (9.49)$$

$$\sigma = \begin{cases} 0.07, & f \le f_p \\ 0.09, & f > f_p \end{cases} \quad (9.50)$$

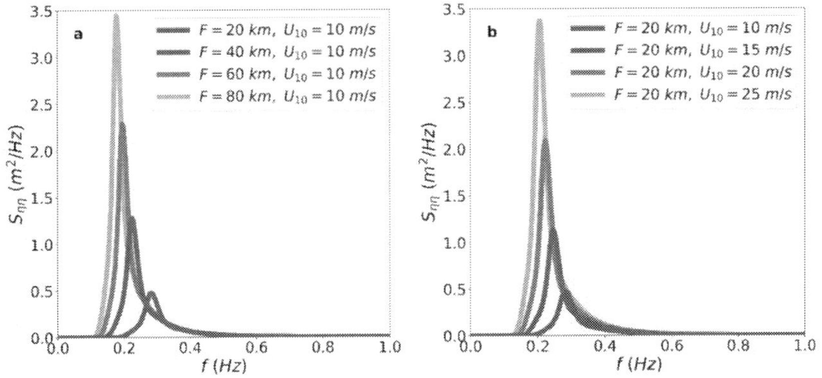

Figure 9.5 Illustration of the JONSWAP spectrum (Hasselmann et al., 1973). Plot (a) shows the JONSWAP spectrum for the wind velocity of $U_{10} = 10\ m/s$, a wind fetch of $F = 20, 40, 60, 80\ km$, and a peak enhancement parameter of $\gamma = 3.3$. Plot (b) shows the JONSWAP spectrum for the wind velocity of $U_{10} = 10, 15, 20, 25\ m/s$, a wind fetch of $F = 20\ km$, and a peak enhancement parameter of $\gamma = 3.3$.

where, f is a frequency in Hz, the f_p is a peak wave frequency in Hz, the α is the Phillip's constant, F is a wind fetch, $\hat{F} = gF/U_{10}^2$ is a dimensionless wind fetch, g is the gravitational acceleration, U_{10} is a 10-min averaged wind velocity at 10 m height from a surface, σ is a spectral width parameter, γ is a peak enhancement parameter. The γ value ranges between 1 and 7 with an average value of $\gamma = 3.3$. The γ represents a ratio of the JONSWAP spectrum over Pierson and Moskowitz spectrum at a peak frequency, i.e. $S_{\eta\eta}(f_p)$. For $\gamma = 1$, the JONSWAP spectrum becomes the same as the Pierson and Moskowitz spectrum. Mitsuyasu et al. (1980) suggested a relationship for γ as:

$$\gamma = 7\left(\frac{gF}{U_{10}^2}\right)^{(-1/7)} = 4.42\left(\frac{f_p U_{10}}{g}\right)^{(3/7)} \tag{9.51}$$

9.7.4 Donelan (1985) spectrum for deep water

Donelan, et al. (1985) proposed a modified form of the JONSWAP spectrum as:

$$S_{\eta\eta}(f) = \frac{\alpha g^2}{(2\pi)^4 f^4 f_p} \exp\left(-\left(\frac{f_p}{f}\right)^4\right) \gamma^{\exp\left(-\frac{(f-f_p)^2}{2\sigma^2 f_p^2}\right)} \tag{9.52}$$

$$\alpha = 0.006\left(\frac{U_{10}}{C_p}\right)^{0.55} \tag{9.53}$$

$$\gamma = \begin{cases} 1.7, & 0.83 < \frac{U_{10}}{C_p} < 1 \\ 1.7 + 6\log_{10}\left(\frac{U_{10}}{C_p}\right), & 1 \le \frac{U_{10}}{C_p} < 5 \end{cases} \tag{9.54}$$

$$\sigma = 0.08\left(1 + \frac{4}{\left(\frac{U_{10}}{C_p}\right)^3}\right) \quad \text{for} \quad 0.83 < \frac{U_{10}}{C_p} < 5 \tag{9.55}$$

where, C_p is a peak wave celerity, which in deep water is $C_p = g/(2\pi f_p)$, and U_{10}/C_p is an inverse wave age. Values of α, γ , and σ in the term of the peak wave frequency, f_p, are (Wiberg and Sherwood, 2008):

$$\alpha = 0.0165\left(\frac{f_p U_{10}}{g}\right)^{0.55} \tag{9.56}$$

$$\gamma = \begin{cases} 1.7, & \frac{f_p U_{10}}{g} < 0.16 \\ 6.5 + 2.6\ln\left(\frac{f_p U_{10}}{g}\right), & \frac{f_p U_{10}}{g} \ge 0.16 \end{cases} \tag{9.57}$$

$$\sigma = 0.08 + 0.0013\left(\frac{f_p U_{10}}{g}\right)^{-3} \tag{9.58}$$

9.7.5 TMA (1985) spectrum in depth-limited water

The most well-known water surface elevation power spectral density for depth-limited water is the TMA spectrum, proposed by Bouws et al., (1985). It is obtained by transforming the JONSWAP spectrum in deep water to the TMA spectrum in shallow water as (Hughes, 1984 and Bouws et al., 1985):

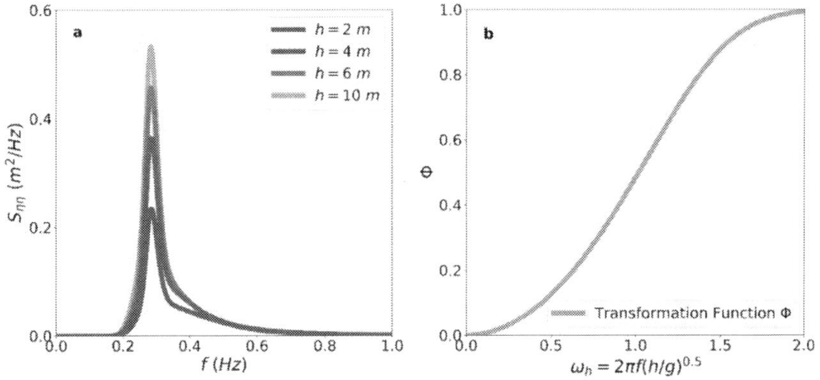

Figure 9.6 Illustration of the TMA spectrum (Bouws et al., 1985). Plot (a) shows the TMA spectrum for the wind velocity of $U_{10} = 10\ m/s$, a wind fetch of $F = 20\ km$, and water depth of $h = 2, 4, 6, 10\ m$. Plot (b) shows a transformation function, Φ, from the JONSWAP spectrum in deep water to the TMA spectrum in depth-limited water.

$$S_{\eta\eta}(f) = \frac{\alpha g^2}{(2\pi)^4 f^5} \exp\left(-1.25\left(\frac{f_p}{f}\right)^4\right) \gamma^{\exp\left(-\frac{(f-f_p)^2}{2\sigma^2 f_p^2}\right)} \Phi$$
$$= S_{\eta\eta_{JONSWAP}} \times \Phi \tag{9.59}$$

$$\alpha = 0.0078 \left(\frac{2\pi U_{10}^2}{g L_p}\right)^{0.49} \tag{9.60}$$

$$\gamma = 2.47 \left(\frac{2\pi U_{10}^2}{g L_p}\right)^{0.39} \tag{9.61}$$

where, L_p is a peak wavelength, which is a wavelength associated with a peak wave frequency.

The Φ is a transformation function. It transforms the JONSWAP spectrum in deep water into the TMA spectrum in depth-limited water as (Kitaigordskii et al., 1975; Hughes, 1984; Bouws et al., 1985; Holthuijsen, 2007):

$$\Phi = \frac{R_{\omega_h}^{-2}}{1 + 2\omega_h^2 R_{\omega_h} / \sinh\left(2\omega_h^2 R_{\omega_h}\right)} \tag{9.62}$$

where, R_{ω_h} and ω_h are:

$$R_{\omega_h} \tanh(\omega_h^2 R_{\omega_h}) = 1 \tag{9.63}$$

$$\omega_h = 2\pi f \sqrt{\frac{h}{g}} \tag{9.64}$$

where, h is water depth.

A value of Φ can be calculated from (Tucker, 1994; Holthuijsen, 2007):

$$\Phi = \frac{\tanh^2(kh)}{1 + 2kh/\sinh(2kh)} \tag{9.65}$$

where, k is a wavenumber and h is water depth.

Also, A value of Φ can be approximated by (Thompson and Vincent, 1983; Bergdahl, 2009):

$$\Phi(f,h) \approx \begin{cases} \frac{\omega_h^2}{2}, & \omega_h \leq 1 \\ 1 - 0.5(2-\omega_h)^2, & 1 < \omega_h < 2 \\ 1, & \omega_h \geq 2 \end{cases} \tag{9.66}$$

Young and Verhagen (1996b) and Young and Babanin (2006) discussed a wave spectrum in a depth-limited condition based on data collected at Lake George, Australia.

There are other wave spectra in the literature (e.g. Liu et al., 2017), such as the ones suggested by Toba ISSC, (1964), International Towing Tank Conference, ITCC (Mathews, 1972; Stansberg et al., 2002), Mitsuyasu (1973), Toba (1973), Ochi and Hubble (1977), Torsethaugen and Haver (2004).

9.7.6 Generating wave power spectra numerically

To generate a water surface elevation power spectral density from any given parametric spectrum, first, a sampling frequency, f_s, needs to be defined. A generated wave spectrum will have a frequency range of $0 \leq f \leq f_s/2$. Then, a total number of the points that represent a frequency range of $0 \leq f \leq f_s - df$, is chosen to be N, where $df = f_s/N$. Assuming N is an even number, then the spectrum, which has a frequency range of $0 \leq f \leq f_s/2$, would consist of $(N/2)+1$ data points.

As an example, assume a sampling frequency of the data is chosen as $f_s = 4\ Hz$. Therefore, a generated spectrum has a frequency range of $0 \leq f \leq 2\ Hz$. If a total number of points is chosen to be $N = 2^8 = 256$, then a generated spectrum has $(256/2)+1 = 129$ data points within a frequency range of $0 \leq f \leq 2\ Hz$. In other words, this spectrum represents a time series dataset which has a sampling frequency of $f_s = 4\ Hz$, a total number of points of $N = 256$, and total duration of $N/f_s = 256/4 = 64\ s$.

9.8 Parametric wave models

As described earlier, using parametric wave spectra is one way to represent water wave properties. A water surface elevation power spectral density, $S_{\eta\eta}$, represents properties of the random sea within a range of frequencies. Another method to estimate wave properties in water bodies such as an ocean and a sea is to use parametric wave models. Unlike wave spectrum, parametric wave models only predict representative wave height and wave period, and do not provide any information on a distribution of wave properties in different frequencies. Parametric wave models are mostly developed empirically or theoretically based on laboratory and field measured data. These models predict wave properties from wind data and physical properties of the water body.

Most parametric wave models give results either in the form of non-dimensional wave energy, $\hat{E} = g^2 m_0/U_{10}^4$, and non-dimensional wave frequency, $\widehat{f_p} = f_p U_{10}/g$, or in the form of non-dimensional wave height, such as gH_{m0}/U_{10}^2 or gH_s/U_{10}^2, and non-dimensional wave period, such as gT_p/U_{10} or gT_s/U_{10}. In case that results are presented in a form of non-dimensional wave energy and non-dimensional wave frequency, zero-moment wave height, H_{m0}, and a peak wave period, T_p, are calculated as:

$$H_{m0} = 4\sqrt{m_0} = 4\sqrt{\frac{\hat{E}U_{10}^4}{g^2}} \quad (9.67)$$

$$T_p = \frac{1}{f_p} = \frac{1}{(\hat{f_p} g / U_{10})} \tag{9.68}$$

where, $\hat{E}$ is dimensionless wave energy, $\widehat{f_p}$ is a dimensionless peak wave frequency, m_0 is the zero-moment of the water surface elevation power spectral density, $f_p$is a peak wave frequency in Hz, g is the gravitational acceleration, U_{10} is a 10-min averaged wind velocity at 10 m height from a surface, H_{m0} is zero-moment wave height, T_p is a peak wave period, H_s is significant wave height, and T_s is a significant wave period.

9.8.1 Fetch-limited wave growth, duration-limited wave growth and equivalent wind fetch

A wind fetch is one of the main parameters used in parametric wave calculations. To have a correct wave prediction, it is important that wind has enough sustained duration to let a wave travels a length of the full fetch. In other words, predicted wave properties by parametric models are reliable only if a duration of the sustained wind in the direction of the wind fetch is long enough to provide sufficient time for wind waves to travel a full fetch and reach a destination.

If a wind duration is long enough to allow wind waves travel a full length of the wind fetch, then wind wave growth is fetch-limited. However, if sustained wind duration is not long enough to allow wind waves travel a full fetch, then wave growth is duration-limited (e.g. Karimpour et al., 2017).

To define if wave growth is fetch-limited or duration-limited, first, a duration of the sustained wind is calculated. Then, a minimum duration of the sustained wind that allows wind waves to travel a full fetch is calculated. There are several equations in the literature that can be used to calculate a minimum required duration for a wave to travel a full fetch. After a minimum required duration is calculated, it is compared with sustained wind duration. If a duration of the sustained wind is equal or larger than a minimum required duration, then wave growth is fetch-limited. Otherwise, if a duration of the sustained wind is smaller than a minimum required duration, then wave growth is duration-limited.

In case that wave growth is duration-limited, an equivalent fetch needs to be calculated. An equivalent fetch is equal to a distance that wind waves can travel during a duration of the sustained wind. Some parametric wave models have their own formula to calculate an equivalent fetch for a duration of the sustained wind. However, if such a formula is not provided by a model, an equivalent fetch can be obtained from an equation that is used to calculate a minimum required duration. To do that, an equation for calculating a minimum required duration is re-arranged to present a wind fetch as a function of a minimum required duration. Then a minimum required time in the equation is replaced by a sustained wind duration and the equation is solved to get an equivalent fetch.

For a wave calculation, if a sustained wind duration is long enough and wave growth is fetch-limited, then full fetch values are used in a calculation. Otherwise, if a sustained wind duration is not long enough to let a wind wave travels a full fetch and wave growth becomes duration-limited, then an equivalent fetch instead of a full fetch, based on situations, might need to be used in a calculation.

9.8.2 Fully developed condition

A fully developed condition refers to a condition which a wave does not grow any further. In other words, in a fully developed condition, wave height (wave energy) and a wave period reach their largest value and do not become larger. Similarly, a peak wave period reaches its largest value and does not become larger in a fully developed condition. It means that a peak wave frequency reaches its smallest value and does not become smaller under this situation. One of the most well-known relationships that describe a fully developed condition for wave growth is proposed by Pierson and Moskowitz (1964) as:

$$\hat{E}_{FullyDev} = 3.64 \times 10^{-3} \tag{9.69}$$

$$\widehat{f_p}_{FullyDev} = 0.133 \tag{9.70}$$

where, $\hat{E}_{FullyDev}$ is a non-dimensional value for fully developed wave energy (the largest value of the wave energy), and $\widehat{f_p}_{FullyDev}$ is a non-dimensional value for a fully developed peak wave frequency (the smallest value of the peak wave frequency). All of the predicted wave values by parametric models should always be limited to these two boundaries as $\hat{E} \leq \hat{E}_{FullyDev}$ and $\widehat{f_p} \geq \widehat{f_p}_{FullyDev}$.

9.8.3 Asymptotic limit

In shallow and intermediate water, or in so-called depth-limited water, water depth limits the growth of the wind-generated waves. This limitation of the wave growth in shallow and intermediate water is called an asymptotic limit. For any given depth, wave height (wave energy) and a wave period reach their largest value and do not become larger. Similarly, for the same depth, a peak wave period reaches its largest value and does not become larger. It means that a peak wave frequency reaches its smallest value and does not become smaller under this situation. These ultimate values are known as asymptotic limits of the wave growth in shallow and intermediate water. All of the predicted wave values by parametric models in shallow and intermediate water should always be limited to asymptotic limits as $\hat{E} \leq \hat{E}_{Asympt}$ and $\widehat{f_p} \geq \widehat{f_p}_{Asympt}$, where, $\hat{E}_{Asympt}$ is a non-dimensional asymptotic limit for wave energy, and $\widehat{f_p}_{Asympt}$ is a non-dimensional asymptotic limit for a peak wave frequency. Asymptotic limit values that are applicable in depth-limited water should not be mistaken for a fully develop values that are applicable to all water depth, particularly in deep water. To find values of asymptotic limits, refer to parametric wave models in shallow and intermediate water.

9.8.4 Mean water depth along a wind fetch

To use parametric wave models in shallow and intermediate water, or in so-called depth-limited water, water depth values are required. Parametric models generally use mean water depth along a wind fetch rather than local water depth. It means that water depth should be averaged along a wind fetch before being used in a model. For these models, mean water depth is averaged along a wind fetch as:

$$\bar{h} = \frac{\int_0^F h(x)dx}{F} \approx \frac{\sum_i h_i \Delta x_i}{\sum_i \Delta x_i} \tag{9.71}$$

where, $\bar{h}$ is mean water depth averaged along a wind fetch, $h(x)$ is spatial water depth along a wind fetch axis, x is a longitudinal axis along a wind fetch, and F is a wind fetch. If a water body has a horizontal bathymetry, i.e. horizontal bottom, then $\bar{h} = h$, where h is local water depth at a study point.

After mean water depth along a wind fetch is obtained, dimensionless water depth is calculated as:

$$\hat{h} = \frac{g\bar{h}}{U_{10}^2} \tag{9.72}$$

where, $\hat{h}$ is non-dimensional water depth based on mean water depth, $\bar{h}$.

9.8.5 Steps to calculate wave properties using parametric models

Here are steps to calculate wave properties from parametric models. More or less, these steps can be used for any parametric model.

1. Obtain wind data, i.e. velocity magnitude and velocity direction. Usually, these can be attained from a nearby weather station.
2. Convert obtained wind data to the wind data that match wave model requirements. Commonly, wave models use 10-min averaged wind velocity and wind direction at 10 m height above a surface. If a wave model requires otherwise, convert wind data to match model requirements.
3. Obtain bathymetry data for a water body that is subjected to the study.
4. Define a boundary of the water body.
5. Define a series of wind directions between 0° to 360° at specific intervals, such as at every 5°. Then, find a wind fetch from a boundary to the point of interest for all these directions. At each direction, a wind fetch is calculated by drawing a line from a water

boundary toward a point of interest with the same direction as a wind direction. A length of this line is a wind fetch for that wind direction.

6. Use wind fetch data from a previous step and interpolate wind fetches for all wind directions in a dataset.
7. If a water body has a shallow depth, then water depth needs to be defined. Most of the wave models which are developed for shallow and intermediate water use depth value that is averaged along a wind fetch. Therefore, average depth along a wind fetch for all wind directions in a dataset needs to be defined. Similar to a wind fetch, first, averaged depth along a wind fetch is calculated at specific directional intervals. Then, depth values are interpolated for all wind directions. For an averaging procedure, if depth does not vary a lot, then use a simple averaging, otherwise, use an integral averaging as $\bar{h} = \int_0^F h(x)dx/F \approx \sum_i h_i \Delta x_i / \sum_i \Delta x_i$.
8. Calculate a sustained wind duration. Wind is considered to be sustained when both wind velocity magnitude and wind velocity direction are steady. Coastal Engineering Manual, CEM (2015), considers wind to be steady if a wind velocity magnitude does not deviate more than 2.5 m/s from its mean value and at the same time, a wind velocity direction does not deviate more than 15° from its mean value.
9. Calculate a minimum duration of the sustained wind that is required for a wave to travel a full length of the wind fetch and becomes fetch-limited.
10. Compare a sustained wind duration with a minimum duration of the sustained wind required for a wave to become fetch-limited. If a sustained wind duration is equal or larger than a minimum required duration, then wave growth is fetch-limited and a full fetch can be used in a calculation. Otherwise, if a sustained wind duration is smaller than a minimum required duration, then wave growth is a duration-limited. In such case, an equivalent wind fetch (depending on the conditions) might need to be calculated and be used in a parametric wave model.
11. If a wave model uses dimensionless values, calculate dimensionless input values for a wave model, such as a dimensionless wind fetch. If a water body is shallow, dimensionless depth needs to be calculated as well.

12. Use a parametric wave model to calculate characteristic wave properties such as a dimensionless peak wave frequency (or a dimensionless peak wave period) and dimensionless wave energy (or dimensionless significant wave height or dimensionless zero-moment wave height). Some model might give dimensional values instead of dimensionless values.
13. Define values for a fully developed condition. Compare values obtained from a parametric wave model with a fully developed condition. Values obtained from a parametric wave model should be limited to the values of the fully developed condition.
14. If a water body is shallow, calculate values of asymptotic limits. Compare values obtained from a parametric wave model with asymptotic limits. Values obtained from a parametric wave model should be limited to the values of the asymptotic limits.
15. If values obtained from a parametric wave model are non-dimensional, convert them to dimensional values.

Special attention is required for using an equivalent wind fetch. As it is mentioned above, if a duration of the sustained wind is equal or larger than a minimum required duration for a fetch-limited condition, then wave growth is fetch-limited. Otherwise, if a duration of the sustained wind is smaller than a minimum required duration for a fetch-limited condition, then wave growth is duration-limited. In case that wave growth is duration-limited, an equivalent fetch is calculated for a duration of the sustained wind, and is used instead of the full wind fetch in a parametric wave model. However, in cases that a wind direction changes rapidly, the wind never becomes sustained. In these conditions, comparing a sustained wind duration with a minimum required duration of the sustained wind for fetch-limited condition always leads to a duration-limited condition. This means that an equivalent fetch needs to be used instead of a full fetch in all calculations. However, using an equivalent fetch when a wind direction varies rapidly can lead to underestimation of the wave energy as it ignores a pre-existing wave energy in a domain. Therefore, for these conditions, although a fetch-limited condition for wave growth is not met, a full fetch needs to be used in a parametric wave model to get reasonable results.

9.9 Parametric wave model equations

There are a series of parametric wave models in the literature for wave predictions in shallow, intermediate and deep water. Here, some of the most prominent parametric wave models in shallow, intermediate and deep water are presented.

9.9.1 SMB parametric wave model in deep water

One of the earliest wave prediction methods in deep water is the SMB method. It is named after the works of Sverdrup and Munk (1947) and Bretschneider (1952, 1958). The SMB method suggested wave growth equations in deep water as:

$$\frac{gH_s}{U_{10}^2} = 0.283 \tanh\left(0.0125\hat{F}^{0.42}\right) \tag{9.73}$$

$$\frac{gT_s}{U_{10}} = 7.54 \tanh\left(0.077\hat{F}^{0.25}\right) \tag{9.74}$$

where, H_s, is significant wave height, T_s is a significant wave period ($T_s \approx 0.95T_p$; Shore Protection Manual, SPM, 1984), and $\hat{F}$ is a dimensionless wind fetch. Wave height and a wave period obtained from these equations should be limited to values for a fully developed condition.

The SMB method suggested that a minimum duration of the sustained wind which allows wind waves to travel a full length of the wind fetch and become fetch-limited in deep water is (e.g., Bishop, C. T. 1983; Etemad-Shahidi et al., 2009):

$$\hat{t}_{min} = 6.5882 \exp\left(\left(0.0161\left(\ln \hat{F}\right)^2 - 0.3692 \ln \hat{F} + 2.2024\right)^{1/2} + 0.8798 \ln \hat{F}\right) \tag{9.75}$$

where, $\hat{t}_{min} = gt_{min}/U_{10}$ is a non-dimensional minimum duration (of the sustained wind) that is required for a wave to travel a full fetch distance of F, the t_{min}, in second, is a minimum duration (of the sustained wind) that is required for a wave to travel a full fetch distance of F, the $\hat{F} = gF/U_{10}^2$ is a non-dimensional wind fetch, the F is a wind fetch, g is the gravitational acceleration, and U_{10} is a 10-min averaged wind velocity at 10 m height from a surface.

The SMB method suggested asymptotic limits for wave growth as (e.g. Young and Verhagen, 1996a):

$$\hat{E}_{Asypmt} = 1.4 \times 10^{-3} \hat{h}^{1.5} \tag{9.76}$$

$$\widehat{f_p}_{Asympt} = 0.16 \hat{h}^{-0.375} \tag{9.77}$$

$$\hat{h} = \frac{g\bar{h}}{U_{10}^2} \tag{9.78}$$

where, $\hat{h} = g\bar{h}/U_{10}^2$ is dimensionless water depth, and $\bar{h}$ is mean water depth along a wind fetch. Note that these asymptotic limits are only used in shallow and intermediate water. Since the SMB method was developed and used in deep water, these asymptotic limits are not required to be used with the SMB method. However, these asymptotic limits might be used in other depth-limited wave models, if they do not suggest any particular asymptotic limits.

It is suggested to use an effective wind fetch, F_{eff}, in the SMB method as described by Saville (1954). To do that, an upwind side of the study point, from $-45°$ to $+45°$ of the wind direction, is divided into radials with an interval of 6°. Then, a wind fetch is averaged as (e.g., Smith, 1991; CIRIA, 2007; Etemad-Shahidi et al., 2009):

$$F_{eff} = \frac{\sum_i F_i \cos\theta_i}{\sum_i \cos\theta_i} \tag{9.79}$$

where, F_{eff} is an effective wind fetch based on Saville (1954), θ_i is a direction of radial with respect to a wind direction within a range of $-45°$ to 45° of the wind direction and an interval of 6°, the F_i is a wind fetch at a radial with a direction of θ_i.

9.9.2 Wilson (1965) parametric wave model in deep water

Wilson (1965) suggested wave growth equations in deep water as (e.g. Goda, 2003):

$$\frac{gH_s}{U_{10}^2} = 0.3\left(1 - \left(1 + 0.004\hat{F}^{(1/2)}\right)^{-2}\right) \tag{9.80}$$

$$\frac{gT_s}{U_{10}} = 1.37 \times 2\pi \times \left(1 - \left(1 + 0.008\hat{F}^{(1/3)}\right)^{-5}\right) \tag{9.81}$$

where, H_s, is significant wave height, and T_s is a significant wave period ($T_s \approx 0.95T_p$; Shore Protection Manual, SPM, 1984), $\hat{F}$ is a dimensionless wind fetch. Wave height and a wave period obtained from these equations should be limited to values for a fully developed condition. A fully developed condition for wave growth proposed by Pierson and Moskowitz (1964) can be re-written as:

$$\left(\frac{gH_{mo}}{U_{10}^2}\right)_{FullyDev} = 4 \times \sqrt{3.64 \times 10^{-3}} \tag{9.82}$$

$$\left(\frac{gT_p}{U_{10}}\right)_{FullyDev} = \frac{1}{0.133} \tag{9.83}$$

$$\left(\frac{gT_s}{U_{10}}\right)_{FullyDev} = \frac{0.95}{0.133} \tag{9.84}$$

Wilson (1965) suggested that a minimum duration of the sustained wind which allows wind waves to travel a full length of the wind fetch and become fetch-limited in deep water is (e.g., Etemad-Shahidi et al., 2009):

$$t_{min} = 43\frac{F^{0.73}}{U_{10}^{0.46}g^{0.27}} \tag{9.85}$$

This relationship can be re-written in a non-dimensional form as:

$$\hat{t}_{min} = 43\hat{F}^{0.73} \tag{9.86}$$

where, $\hat{t}_{min} = gt_{min}/U_{10}$ and $\hat{F} = gF/U_{10}^2$.

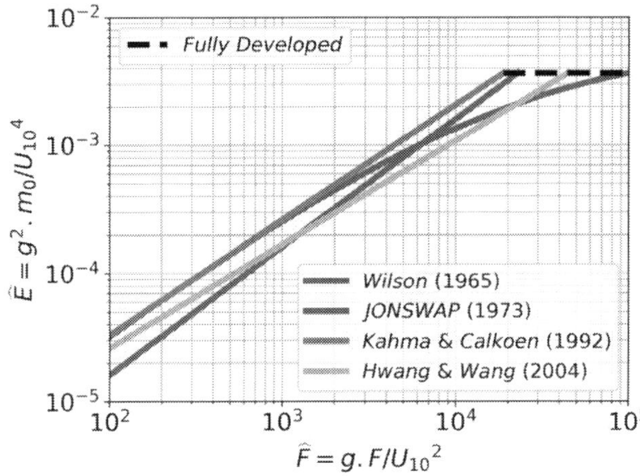

Figure 9.7 Parametric wave models for predicting dimensionless wave energy, $\hat{E} = g^2 m_0 / U_{10}^4$, in deep water. Dimensionless wave energy values for Wilson (1965) calculated based on $H_{m0} \approx H_s$. The value for a fully developed condition is based on Pierson and Moskowitz (1964) as $\hat{E}_{FullyDev} = 3.64 \times 10^{-3}$. Values of $\hat{E}$ are limited to the value of the fully developed condition.

9.9.3 JONSWAP (1973) parametric wave model in deep water

One of the most well-known and widely used parametric wave models in deep water is the JONSWAP wave model. Hasselmann et al., (1973) developed the JONSWAP wave growth equations for deep water during the Joint North Sea Wave Project (JONSWAP) as:

$$\hat{E} = 1.6 \times 10^{-7} \hat{F} \tag{9.87}$$

$$\hat{f_p} = 3.5 \hat{F}^{-0.33} \tag{9.88}$$

where, $\hat{E}$ is dimensionless wave energy, $\hat{f_p}$ is a dimensionless peak wave frequency, and $\hat{F}$ is a dimensionless wind fetch. Wave energy and a peak wave frequency obtained from these equations should be limited to values for a fully developed condition.

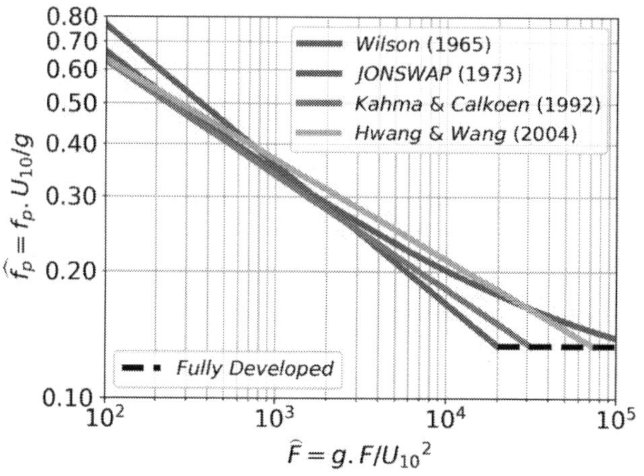

Figure 9.8 Parametric wave models for predicting a dimensionless peak wave frequency, $\widehat{f_p} = f_p U_{10}/g$, in deep water. Dimensionless peak wave frequency values for Wilson (1965) calculated based on $T_s \approx 0.95T_p$ (Shore Protection Manual, SPM, 1984). The value for a fully developed condition is based on Pierson and Moskowitz (1964) as $\widehat{fp}_{FullyDev} = 0.133$. Values of $\widehat{f_p}$ are limited to the value of the fully developed condition.

Hasselmann et al., (1973), suggested a minimum duration of the sustained wind which allows wind waves to travel a full length of the wind fetch and become fetch-limited in deep water as (e.g., Bishop, 1983):

$$t_{min} = 65.6\frac{F^{0.67}}{U_{10}^{0.33}g^{0.33}} \tag{9.89}$$

This relationship can be re-written, with a small approximation, in a non-dimensional form as:

$$\hat{t}_{min} \approx 65.6\hat{F}^{0.67} \approx 65.6\hat{F}^{(2/3)} \tag{9.90}$$

where, $\hat{t}_{min} = gt_{min}/U_{10}$ and $\hat{F} = gF/U_{10}^2$.

Based on the JONSWAP results, Carter (1982) suggested a minimum duration of the sustained wind which allows wind waves to travel a full length of the wind fetch and become fetch-limited in deep water as:

$$t_{min-Hr} = 1.167 \frac{F_{Km}^{0.7}}{U_{10}^{0.4}} \tag{9.91}$$

where, t_{min-Hr} is a minimum duration (of the sustained wind) required for a wave to travel over a wind fetch in *hour*, and F_{km} is a wind fetch in *kilometer*.

9.9.4 Shore Protection Manual, SPM, (1984), parametric wave model in deep water

Based on the JONSWAP model (Hasselmann et al., 1973), the Shore Protection Manual, SPM, (1984), suggested wave growth equations in deep water as:

$$\frac{gH_{m0}}{U_A^2} = 1.6 \times 10^{-3} \left(\frac{gF}{U_A^2}\right)^{1/2} \tag{9.92}$$

$$\frac{gT_p}{U_A} = 2.857 \times 10^{-1} \left(\frac{gF}{U_A^2}\right)^{1/3} \tag{9.93}$$

$$U_A = 0.71 U_{10}^{1.23} \tag{9.94}$$

where, U_A is the adjusted wind velocity, H_{m0} is zero-moment wave height, and T_p is a peak wave period. Wave height and a wave period obtained from these equations should be limited to values for a fully developed condition.

SPM (1984) suggested that a minimum duration of the sustained wind which allows wind waves to travel a full length of the wind fetch and become fetch limited in deep water is:

$$\frac{gt_{min}}{U_A} = 6.88 \times 10^{1} \left(\frac{gF}{U_A^2}\right)^{2/3} \tag{9.95}$$

where t_{min}, in second, is a minimum required time for a wave to travel a fetch distance of F.

SPM (1984) suggested a fully developed condition as:

$$\left(\frac{gH_{m0}}{U_A^2}\right)_{FullyDev} = 2.433 \times 10^{-1} \tag{9.96}$$

$$\left(\frac{gT_p}{U_A}\right)_{FullyDev} = 8.134 \tag{9.97}$$

$$\left(\frac{gt_{min}}{U_A}\right)_{FullyDev} = 7.15 \times 10^4 \tag{9.98}$$

It should be pointed out that U_{10} in SPM (1984) equations is wind velocity at 10 m height from a surface. However, SPM (1984) requires that wind velocity to be adjusted based on a sustained wind duration. To do that, wind velocity values are recalculated by using a gust factor proposed by SPM (1984) to adjust a wind averaging period to a sustained wind duration. For example, if U_{10} is representing a 10-min averaged wind velocity, and a sustained wind duration is 15 minutes, then velocity values should be converted from 10-min averaged wind to 15-min averaged wind by using a gust factor.

9.9.5 Kahma and Calkoen (1992) parametric wave model in deep water

Kahma and Calkoen (1992) suggested wave growth equations in deep water as (e.g., Karimpour and Chen, 2016):

$$\hat{E} = 5.2 \times 10^{-7} \hat{F}^{0.9} \tag{9.99}$$

$$\hat{f}_p = \frac{13.7}{2\pi} \hat{F}^{-0.27} \tag{9.100}$$

Wave energy and a peak wave frequency obtained from these equations should be limited to values for a fully developed condition.

9.9.6 Hwang and Wang (2004) parametric wave model in deep water

Hwang and Wang (2004) suggested wave growth equations in deep water as (e.g., Karimpour and Chen, 2016):

$$\hat{E} = 6.191 \times 10^{-7} \hat{F}^{0.8106} \tag{9.101}$$

$$\hat{f}_p = \frac{11.86}{2\pi} \hat{F}^{-0.2368} \tag{9.102}$$

Wave energy and a peak wave frequency obtained from these equations should be limited to values for a fully developed condition.

9.9.7 Coastal Engineering Manual, CEM (2015), parametric wave model in deep water

The Coastal Engineering Manual, CEM (2015), suggested wave growth equations in deep water as:

$$\frac{gH_{m0}}{u_*^2} = 4.13 \times 10^{-2} \left(\frac{gF}{u_*^2}\right)^{1/2} \tag{9.103}$$

$$\frac{gT_p}{u_*} = 0.651 \left(\frac{gF}{u_*^2}\right)^{1/3} \tag{9.104}$$

where, u_* is friction velocity (a so-called shear velocity),. and C_D is a drag coefficient. Based on the Coastal Engineering Manual, CEM (2015), u_* and C_D are calculated as:

$$u_*^2 = C_D U_{10}^2 \tag{9.105}$$

$$C_D = 0.001(1.1 + 0.035U_{10}) \tag{9.106}$$

Wave height and a wave period obtained from these equations should be limited to values for a fully developed condition.

CEM (2015) suggested that a minimum duration of the sustained wind which allows wind waves to travel a full length of the wind fetch and become fetch-limited in deep water is:

$$t_{min} = 77.23 \frac{F^{0.67}}{U_{10}^{0.34} g^{0.33}} \tag{9.107}$$

This relationship can be re-written, with a small approximation, in a non-dimensional form as:

$$\hat{t}_{min} = 77.23\hat{F}^{0.67} \approx 77.23\hat{F}^{(2/3)} \quad (9.108)$$

where, t_{min}, in second, is a minimum duration (of the sustained wind) that is required for a wave to travel a full fetch distance of F, the $\hat{t}_{min} = gt_{min}/U_{10}$ is a non-dimensional minimum duration (of the sustained wind) that is required for a wave to travel a full fetch distance of F, the $\hat{F} = gF/U_{10}^2$ is a non-dimensional wind fetch, the F is a wind fetch, g is the gravitational acceleration, and U_{10} is a 10-min averaged wind velocity at 10 m height from a surface.

In case of the duration-limited condition, an equivalent fetch is calculated based on a sustained wind duration as:

$$\frac{gF_{eqv}}{u_*^2} = 5.23 \times 10^{-3} \left(\frac{gt_{st}}{u_*}\right)^{(3/2)} \quad (9.109)$$

where, F_{eqv} is an equivalent fetch, and t_{st} is a sustained wind duration in second.

The CEM (2015) suggested a fully developed condition as:

$$\left(\frac{gH_{m0}}{u_*^2}\right)_{FullyDev} = 2.115 \times 10^2 \quad (9.110)$$

$$\left(\frac{gT_p}{u_*}\right)_{FullyDev} = 2.398 \times 10^2 \quad (9.111)$$

It should be pointed out that U_{10} in CEM (2015) equations is wind velocity at 10 m height from a surface. However, CEM (2015) requires that wind velocity to be adjusted based on a sustained wind duration. To do that, wind velocity values are recalculated by using a gust factor proposed by CEM (2015) to adjust a wind averaging period to a sustained wind duration. For example, if U_{10} is representing a 10-min averaged wind velocity, and a sustained wind duration is 15 minutes, then velocity values should be converted from 10-min averaged wind to 15-min averaged wind by using a gust factor.

9.9.8 Shore Protection Manual, SPM, (1984), parametric wave model in depth-limited water

Based on the JONSWAP model (Hasselmann et al., 1973), the Shore Protection Manual, SPM, (1984), suggested wave growth equations in shallow and intermediate water depth as:

$$\frac{gH_{m0}}{U_A^2} = 0.283 \tanh\left(0.53\left(\frac{g\bar{h}}{U_A^2}\right)^{3/4}\right)\tanh\left(\frac{0.00565\left(gF/U_A^2\right)^{1/2}}{\tanh\left(0.53\left(g\bar{h}/U_A^2\right)^{3/4}\right)}\right) \quad (9.112)$$

$$\frac{gT_s}{U_A} = 7.54 \tanh\left(0.833\left(\frac{g\bar{h}}{U_A^2}\right)^{3/8}\right)\tanh\left(\frac{0.0379\left(gF/U_A^2\right)^{1/3}}{\tanh\left(0.833\left(g\bar{h}/U_A^2\right)^{3/8}\right)}\right) \quad (9.113)$$

$$U_A = 0.71U_{10}^{1.23} \quad (9.114)$$

where, U_A is the adjusted wind velocity, H_{m0} is zero-moment wave height, T_s is a significant wave period ($T_s \approx 0.95T_p$; SPM, 1984), the $\bar{h}$ is mean water depth along a wind fetch, and F is a wind fetch. Wave height and a wave period obtained from these equations should be limited to values for a fully developed condition.

SPM (1984) suggested that a minimum duration of the sustained wind which allows wind waves to travel a full length of the wind fetch and become fetch limited in deep water is:

$$\frac{gt_{min}}{U_A} = 5.37 \times 10^2\left(\frac{gT_s}{U_A}\right)^{7/3} \quad (9.115)$$

where, t_{min}, in second, is a minimum required time for a wave to travel a fetch distance of F.

SPM (1984) did not suggest values for asymptotic wave growth limits in shallow and intermediate water. If values obtained from the SPM method are required to be checked by asymptotic limits, alternative equations such as asymptotic limits suggested by SMB method might be used. The SMB method suggested asymptotic wave growth limits as (e.g. Young and Verhagen, 1996a):

$$\hat{E}_{Asypmt} = 1.4 \times 10^{-3} \hat{h}^{1.5} \tag{9.116}$$

$$\widehat{f_p}_{Asympt} = 0.16 \hat{h}^{-0.375} \tag{9.117}$$

where, $\hat{h} = g\bar{h}/U_{10}^2$ is dimensionless water depth, and $\bar{h}$ is mean water depth along a wind fetch.

Outcomes of the wave model proposed by the Shore Protection Manual are in terms of zero-moment wave height and significant wave period. However, asymptotic values obtained from The SMB method are in terms of dimensionless energy and dimensionless peak wave frequency. Therefore, asymptotic values of the dimensionless energy, $\hat{E}_{Asypmt}$, and dimensionless peak wave frequency, $\widehat{f_p}_{Asympt}$, should be converted to asymptotic values of the zero-moment wave height and significant wave period before they can be compared with SPM results. To do that, first, asymptotic limits proposed by SMB method based on U_{10} are re-written as:

$$\left(\frac{gH_{m0}}{U_{10}^2}\right)_{FullyDev} = 4 \times \sqrt{\hat{E}_{Asypmt}} \tag{9.118}$$

$$\left(\frac{gT_p}{U_{10}}\right)_{FullyDev} = \frac{1}{\widehat{f_p}_{Asympt}} \tag{9.119}$$

$$\left(\frac{gT_s}{U_{10}}\right)_{FullyDev} = \frac{0.95}{\widehat{f_p}_{Asympt}} \tag{9.120}$$

Then, these values are recalculated based on U_A. It is because that the asymptotic limit equations by SMB use U_{10}, while SPM equations use $U_A = 0.71U_{10}^{1.23}$. Therefore, these values should be recalculated based on U_A before being compared with SPM values.

SPM (1984) suggested a fully developed condition as:

$$\left(\frac{gH_{m0}}{U_A^2}\right)_{FullyDev} = 2.433 \times 10^{-1} \tag{9.121}$$

$$\left(\frac{gT_p}{U_A}\right)_{FullyDev} = 8.134 \tag{9.122}$$

$$\left(\frac{gt_{min}}{U_A}\right)_{FullyDev} = 7.15 \times 10^4 \tag{9.123}$$

Then, by considering $T_s \approx 0.95T_p$ (SPM, 1984), a fully developed condition for a significant wave period, T_s, can be re-written as:

$$\left(\frac{gT_s}{U_A}\right)_{FullyDev} = 0.95 \times 8.134 \tag{9.124}$$

It should be pointed out that U_{10} in SPM (1984) equations is wind velocity at 10 m height from a surface. However, SPM (1984) requires that wind velocity to be adjusted based on a sustained wind duration. To do that, wind velocity values are recalculated by using a gust factor proposed by SPM (1984) to adjust a wind averaging period to a sustained wind duration. For example, if U_{10} is representing a 10-min averaged wind velocity, and a sustained wind duration is 15 minutes, then velocity values should be converted from 10-min averaged wind to 15-min averaged wind by using a gust factor.

9.9.9 Young and Verhagen (1996a) parametric wave model in depth-limited water

Based on the SPM (1984) equations and data collected at Lake George, Australia, Young and Verhagen (1996a) suggested wave growth equations in shallow and intermediate water depth as:

$$\hat{E} = 0.00364\left(\tanh\left(0.493(\hat{h})^{3/4}\right)\tanh\left(\frac{0.00313(\hat{F})^{0.57}}{\tanh\left(0.493(\hat{h})^{3/4}\right)}\right)\right)^{1.74} \tag{9.125}$$

$$\hat{f}_p = 0.133\left(\tanh\left(0.331(\hat{h})^{1.01}\right)\tanh\left(\frac{0.0005215(\hat{F})^{0.73}}{\tanh\left(0.331(\hat{h})^{1.01}\right)}\right)\right)^{-0.37} \tag{9.126}$$

where, $\hat{F} = gF/U_{10}^2$ is a non-dimensional wind fetch, $\hat{h} = g\bar{h}/U_{10}^2$ is dimensionless water depth, $\bar{h}$ is mean water depth along a wind fetch, F is a wind fetch, g is the gravitational acceleration, and U_{10} is a 10-min averaged wind velocity at 10 m height from a surface. Wave energy and a peak wave frequency obtained from these equations should be limited to values of the asymptotic limits and values of the fully developed condition.

Young and Verhagen (1996a) suggested asymptotic limits for fetch-limited shallow water as:

$$\hat{E}_{Asympt} = 1.06 \times 10^{-3}\hat{h}^{1.3} \tag{9.127}$$

$$\hat{f}_{p_{Asympt}} = 0.2\hat{h}^{-0.375} \tag{9.128}$$

Young and Babanin (2006) modified Young and Verhagen (1996a) asymptotic limit relationship for the wave energy as:

$$\hat{E}_{Asympt} = 1.0 \times 10^{-3}\hat{h}^{1.2} \tag{9.129}$$

9.9.10 Karimpour et al. (2017) parametric wave model in depth-limited water

Based on data collected in Louisiana estuaries, USA, Karimpour et al. (2017) suggested wave growth equations in shallow and intermediate water depth as:

$$\hat{E} = 3.0 \times 10^{-8}\hat{F}^{\left(2.7(\hat{F}/\hat{h})^{-0.1}\right)} \tag{9.130}$$

$$\hat{f}_p = 3.5\hat{F}^{\left(-0.75(\hat{F}/\hat{h})^{-0.1}\right)} \tag{9.131}$$

where, $\hat{F} = gF/U_{10}^2$ is a non-dimensional wind fetch, $\hat{h} = g\bar{h}/U_{10}^2$ is dimensionless water depth based on mean water depth along a wind fetch, $\bar{h} = \frac{1}{F}\int_0^F h(x)dx$ is mean water depth averaged along a wind fetch, $h(x)$ is spatial water depth along a wind fetch axis, x is a longitudinal axis along a wind fetch, F is a wind fetch, g is the gravitational acceleration, and U_{10} is a 10-min averaged wind velocity at 10 m height from a surface.

Wave energy and a peak wave frequency obtained from these equations should be limited to values of the asymptotic limits and values of the fully developed condition. Results from these equations should be limited to the deep-water wind wave growth rates suggested by JONSWAP (Hasselmann et al., 1973), i.e. $\hat{f}_p \geq 3.5\hat{F}^{-0.33}$ and $\hat{E} \leq 1.6 \times 10^{-7}\hat{F}$. Also, values from these equations for a wind fetch $\hat{F} \geq 2 \times 10^4$ are limited to values for a wind fetch

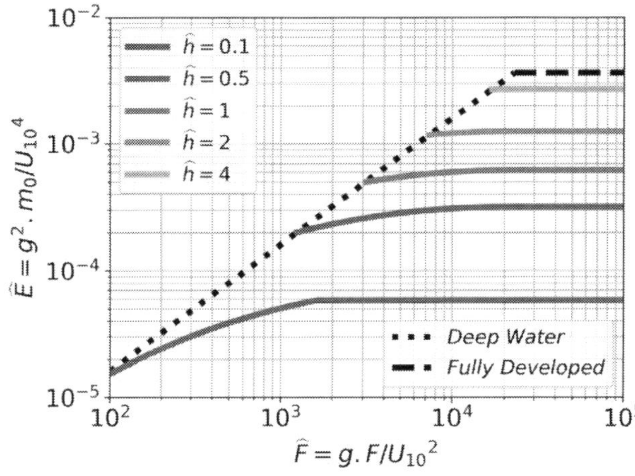

Figure 9.9 A parametric wave model for predicting dimensionless wave energy, $\hat{E} = g^2 m_0 / U_{10}^4$, in shallow and intermediate water depth, based on Karimpour et al. (2017) as $\hat{E} = 3.0 \times 10^{-8} \hat{F}^{\left(2.7(\hat{F}/\hat{h})^{-0.1}\right)}$. Values for a fully developed condition are based on the Pierson and Moskowitz (1964) as $\hat{E}_{FullyDev} = 3.64 \times 10^{-3}$. Values of $\hat{E}$ are limited to values for deep water and fully developed condition.

equal to $\hat{F} = 2 \times 10^4$, i.e. $\hat{f}_p(\hat{F} \geq 2 \times 10^4) = \hat{f}_p(\hat{F} = 2 \times 10^4)$ and $\hat{E}(\hat{F} \geq 2 \times 10^4) = \hat{E}(\hat{F} = 2 \times 10^4)$.

For shallow and intermediate water depth, i.e. depth-limited water, Karimpour et al. (2017) calculated t_{min} from wave group velocity as $t_{min} = F/\bar{C}_g$, where $\bar{C}_g = \frac{1}{F}\int_0^F C_g(x)dx$ is a mean value of wave group velocity, $C_g(x)$ is spatial wave group velocity along a wind fetch axis, and x is a longitudinal axis along a wind fetch. Then, Karimpour et al. (2017) suggested that a minimum duration of the sustained wind which allows wind waves to travel a full length of the wind fetch and become fetch-limited in shallow and intermediate water depth is:

$$\hat{t}_{min} = 2.59 \left(\frac{\hat{F}}{\hat{h}}\right) \hat{h}^{(2/3)} \leq 77.23 \hat{F}^{(2/3)} \tag{9.132}$$

Note that $\hat{t}_{min}$ in depth-limited water should be always limited to the ones from deep water. It means that the upper limit of $\hat{t}_{min}$ in depth-limited water should be limited to a value of $\hat{t}_{min}$ obtained for either deep water or fully

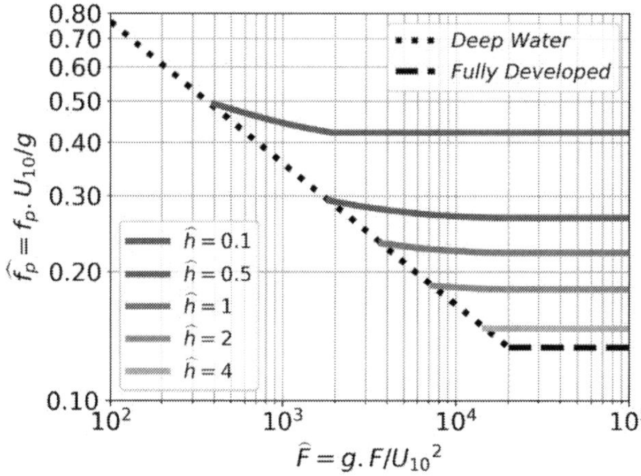

Figure 9.10 A parametric wave model for predicting a dimensionless peak wave frequency, $\widehat{f_p} = f_p U_{10}/g$, in shallow and intermediate water depth, based on Karimpour et al. (2017) as $\widehat{f_p} = 3.5\hat{F}^{\left(-0.75(\hat{F}/\hat{h})^{-0.1}\right)}$. Values for a fully developed condition are based on the Pierson and Moskowitz (1964) as $\widehat{f_p}_{FullyDev} = 0.133$. Values of $\widehat{f_p}$ are limited to values for deep water and fully developed condition.

developed condition, whichever is smaller. For a deep-water limit, an equation suggested by CEM (2015) as $\hat{t}_{min} = 77.23\hat{F}^{0.67} \approx 77.23\hat{F}^{(2/3)}$ can be used. For a fully developed condition limit, an equation suggested by SPM (1984) as $gt_{min}/U_A = 7.15 \times 10^4$ can be used, where $U_A = 0.71U_{10}^{1.23}$ is the adjusted wind velocity.

By rearranging an equation suggested by Karimpour et al. (2017), an equivalent fetch for a duration-limited wave in shallow and intermediate water depth, i.e. depth-limited water, can be calculated as:

$$\hat{F}_{eqv} = \frac{\hat{t}_{st}\hat{h}}{2.59\hat{h}^{(2/3)}} = \frac{1}{2.59}\hat{t}_{st}\hat{h}^{(1/3)} \quad (9.133)$$

where, $\hat{F}_{eqv} = gF_{eqv}/U_{10}^2$ is a dimensionless equivalent wind fetch, and $\hat{t}_{st} = gt_{st}/U_{10}$ is a dimensionless sustained wind duration.

Karimpour et al. (2017) suggested asymptotic limits for fetch-limited shallow and intermediate water depth as:

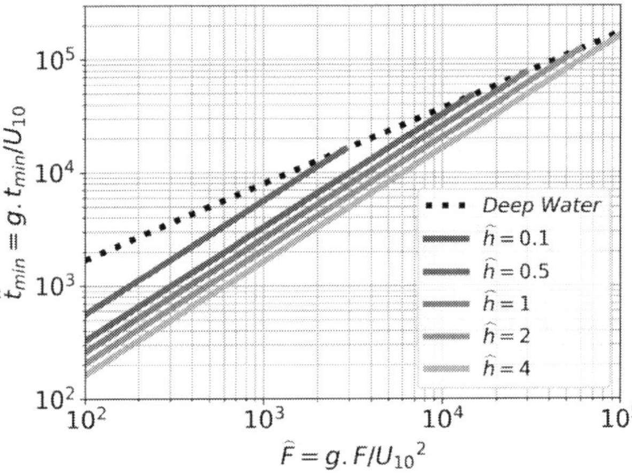

Figure 9.11 A parametric wave model for predicting a dimensionless minimum duration of the sustained wind, $\hat{t}_{min} = gt_{min}/U_{10}$, which is required for a wave to be fetch-limited in shallow and intermediate water depth, based on Karimpour et al. (2017) as $\hat{t}_{min} = 2.59(\hat{F}/\hat{h})\hat{h}^{(2/3)}$. Values for deep water are based on CEM (2015) as $\hat{t}_{min} = 77.23\hat{F}^{0.67}$. Values of $\hat{t}_{min}$ in depth-limited water are limited to deep water values.

$$\hat{E}_{Asympt} = 3 \times 10^{-5} \tan\left(1.56255\big(\tanh(3.356\hat{h})\big)^{0.315}\right) \tag{9.134}$$

$$\hat{f_p}_{Asympt} = 0.133\big(\tanh(0.832\hat{h})\big)^{-0.716}\big(\tanh(2.623\hat{h})\big)^{0.461} \tag{9.135}$$

where, $\hat{E}_{Asympt}$ is a non-dimensional asymptotic limit for wave energy, and $\hat{f_p}_{Asympt}$ is a non-dimensional asymptotic limit for a peak wave frequency. The dimensionless energy and dimensionless peak wave frequency values predicted by a parametric wave model should be limited as $\hat{E} \leq \hat{E}_{Asympt}$ and $\hat{f_p} \geq \hat{f_p}_{Asympt}$.

The asymptotic limits can be re-written in a simpler form as:

$$\hat{f_p}_{Asympt} = \frac{0.7}{1+3\hat{h}^{0.669}} \tag{9.136}$$

$$\hat{E}_{Asympt} = -3.523 \times 10^{-6} + \frac{0.00364}{\left(1+131\big(\exp(\hat{h})\big)^{-6.774}\right)^{0.962}} \tag{9.137}$$

Finally, predicted values by a parametric wave model are compared and limited to a fully developed condition. Pierson and Moskowitz (1964) suggested a fully developed condition for wave growth as $\hat{E}_{FullyDev} = 3.64 \times 10^{-3}$ and $\widehat{f_p}_{FullyDev} = 0.133$, where, $\hat{E}_{FullyDev}$ is dimensionless fully developed wave energy, and $\widehat{f_p}_{FullyDev}$ is a dimensionless fully developed peak wave frequency. The dimensionless energy and dimensionless peak wave frequency values predicted by a parametric wave model should be limited as $\hat{E} \leq \hat{E}_{FullyDev}$ and $\widehat{f_p} \geq \widehat{f_p}_{FullyDev}$.

9.10 Parametric hurricane models

Hurricanes have significant effects on human societies and natural features as their destructive forces cause severe damages and losses every year. Because of their serious effects, hurricanes are subjected to a variety of studies. For that purpose, researchers require having access to some sort of hurricane data such as hurricane's wind and pressure fields. Some researchers may use historical data in their studies, while others might prefer synthetically or numerically generated data. One method to study these natural phenomena is to use parametric models to generate wind and pressure fields of the hurricane and a resultant wave field. Some of these parametric hurricane models are briefly explained here.

9.10.1 Hurricane wind velocity and pressure models

One of the most widely used hurricane models was suggested by Holland (1980). By assuming the Coriolis force is negligible in a maximum wind region, Holland (1980) developed a model for a hurricane wind velocity and air pressure as (e.g., Holland, 1980; Holland et al., 2010):

$$P(r) = P_c + (P_n - P_c)e^{-\left(\frac{R_{mw}}{r}\right)^b} = P_c + \Delta P e^{-\left(\frac{R_{mw}}{r}\right)^b} \tag{9.138}$$

$$V_g(r) = \left(\frac{b(P_n - P_c)\left(\frac{R_{mw}}{r}\right)^b}{\rho_a e^{\left(\frac{R_{mw}}{r}\right)^b}}\right)^{0.5} = \left(\frac{b\Delta P\left(\frac{R_{mw}}{r}\right)^b}{\rho_a e^{\left(\frac{R_{mw}}{r}\right)^b}}\right)^{0.5} \tag{9.139}$$

$$b = \frac{V_{gmax}^2 \times \rho_a \times \exp(1)}{P_n - P_c} \tag{9.140}$$

where, P is air pressure at a surface level at a distance r from a hurricane center in Pa, the V_g is hurricane gradient wind velocity (wind velocity at a gradient level, or mean boundary-layer wind velocity) at a distance r from a hurricane center in (m/s), the P_n is ambient pressure (external pressure) in Pa, the P_c is air pressure at a hurricane center (central pressure) in Pa, the $\Delta P = (P_n - P_c)$ is pressure drop from pressure outside a hurricane, i.e. P_n, to pressure at center of the hurricane, i.e. P_c, in Pa, the r is a radius (distance from a hurricane center), R_{mw} is a radius (distance) to a location of the maximum wind velocity, V_{gmax} is hurricane maximum gradient wind velocity, ρ_a is an air density (Holland, 1980, assumed $\rho_a = 1.15\ kg/m^3$), b is a scaling parameter.

Later, Holland (2008) revised his original model in Holland (1980). Some of the other hurricane wind models that worth mentioning are Rankine model (e.g., Phadke et al., 2003; Holland et al., 2010), model by DeMaria (e.g., DeMaria, 1987; Holland et al., 2010), SLOSH model (Jelesnianski et al., 1992), and model by Emanuel and Rotunno (2011).

Velocity values calculated from parametric models are tangential velocities, which are perpendicular to a radial direction. A rotation direction of the hurricane tangential velocity depends on a location (a latitude) of the hurricane. In the Northern hemisphere, a hurricane rotates counter-clockwise, while in the Southern hemisphere it rotates clockwise.

Values of the velocities and pressure calculated from parametric models should be limited to a maximum radius (outer radius) of a hurricane, where rotational wind velocities become zero and pressure becomes equal to ambient pressure after that. Chavas and Emanuel (2010) suggested that an average value for an outer radius of the hurricane where velocity becomes zero is 423 km (e.g., Lin and Chavas, 2012).

Note that both V_g and V_{gmax} are rotational velocities. Therefore, V_{gmax} should not contain a hurricane translation velocity. In other words, if a value of V_{gmax} contains the effect of the hurricane forward motion, then hurricane translation velocity (also known as forward velocity or background wind) needs to be removed from V_{gmax} before using it to calculate rotational velocity of the hurricane. After rotational velocity of the hurricane is calculated, hurricane translation velocity needs to be added, as a vector, to rotational velocity (Hu et al., 2012; Orton et al, 2016).

9.10.2 Hurricane historical data

To use a parametric hurricane model, values of V_{gmax}, R_{mw}, P_n, P_c, and ρ_a are required. One source for historical hurricane data is U. S. National Hurricane Center (www.nhc.noaa.gov). The National Hurricane Center provides the best track data, HURDAT2, which contains track, wind, pressure, and radius of the previous hurricanes in the Atlantic and North-East and North-Central Pacific (www.nhc.noaa.gov/data).

9.10.3 Hurricane maximum wind velocity

In addition to historical data, there are a variety of relationships to estimate maximum wind velocity, V_{max}, for given ambient pressure, P_n, and hurricane central pressure, P_c. These relationships mostly have a form of $V_{max} = \alpha(P_n - P_c)^{\beta}$, where α and β are empirical constants. One of the most widely used relationships to find V_{gmax} was proposed by Atkinson and Holliday (1977) as:

$$V_{max} = 3.4(1010 - P_c)^{0.644} \tag{9.141}$$

where, V_{max} is 10-minute averaged hurricane maximum wind velocity at 10 m height from a surface, P_c is pressure at a hurricane center (central pressure) in hPa, and $P_n = 1010\ hPa$ is constant ambient pressure, which is valid for the western North Pacific (Holland, 2008).

9.10.4 Hurricane wind velocity adjustment for height and duration

Typically, hurricane wind velocity is given as a 1-min averaged (1-min sustained) gradient wind velocity. However, the most acceptable wind velocity is a 10-min averaged wind velocity at 10 m height from a surface. Therefore, 1-min averaged V_g needs to be converted to 10-min averaged wind velocity at 10 m height from a surface, U_{10} (e.g., Young, 2017). To do that, first, 1-min averaged V_{g-1min} (at a gradient level) is converted to 1-min averaged wind velocity at 10 m height from a surface, $U_{10-1min}$, as:

$$U_{10-1min} = K \times V_{g-1min} \tag{9.142}$$

where, $U_{10-1min}$ is 1-minute averaged wind velocity at 10 m height above a surface, and V_{g-1min} is 1-minute averaged (1-minute sustained) gradient wind velocity (1-minute averaged wind velocity at a gradient level). The K is a conversion factor which is typically reported within a range of 0.7 to 0.8 in the literature (e.g. Shore Protection Manual, 1984; Phadke et al., 2003; Young and Vinoth, 2013; Coastal Engineering Manual, CEM, 2015; Valamanesh et al., 2016; Young, 2017). Some of the proposed values for K in the literature are: $K = 0.8$ (Graham and Numm, 1959; and Powell and Black, 1990), $K = 0.7$ (Harper and Holland, 1999), and $K = 0.71$ (Valamanesh et al., 2016). Note that K values outside of this range are also reported.

After gradient wind velocity is converted to surface wind velocity, 1-min averaged surface wind velocity is converted to 10-min averaged wind velocity as:

$$U_{10} = G \times U_{10-1min} \tag{9.143}$$

where, U_{10} is 10-min averaged wind velocity at 10 m height from a surface, and G is a wind gust factor to convert 1-min averaged wind to 10-min averaged wind. There is a range of values suggested for a gust factor in the literature (e.g. Young, 2017; Liu et al., 2017). Some of the proposed values for G in the literature are: $G = 1/1.1$ (Powell and Houston, 1996), $G = 1/1.08$ (Harper, 2013), $G \approx 1/1.36 \approx 0.88$ (World Meteorological Organization, WMO, 2015), $G = 1/1.08$ to $1/1.16$ (Liu et al., 2017).

9.10.5 Hurricane wind velocity inflow angle

Hurricane velocity calculated by parametric hurricane models is rotational velocity, perpendicular to a radial direction (counter-clockwise in the

Northern hemisphere and clockwise in the Southern hemisphere). In other words, calculated hurricane velocity is tangential velocity. However, in nature, hurricane rotational velocities are not perpendicular to a radial direction and have tendencies toward a center of the hurricane. Therefore, a rotational angle of the hurricane velocity obtained from a parametric model needs to be turned inward to represent a correct rotational angle. To adjust a direction of the hurricane rotational velocity, an inflow angle of the velocity is calculated. One of the widely used relationships for calculating a hurricane velocity inflow angle was proposed by Bretschneider (1972) as (e.g., Phadke et al., 2003; Cheung et al., 2007):

$$\beta(r) = \begin{cases} 10^\circ\left(1 + \frac{r}{R_{mw}}\right), & 0 \le r < R_{mw} \\ 20^\circ + 25^\circ\left(\frac{r}{R_{mw}} - 1\right), & R_{mw} \le r < 1.2R_{mw} \\ 25^\circ, & r \ge 1.2R_{mw} \end{cases} \tag{9.144}$$

where, β is an inflow angle of the hurricane velocity in degree, r is a radius (distance from a hurricane center), R_{mw} is a radius (distance to a location) of the maximum wind velocity. An angle of β is measured inward from an isobar. In other words, β is measured from a tangential direction toward a hurricane center. Note that a hurricane rotates counter-clockwise in the Northern hemisphere and rotates clockwise in the Southern hemisphere. Therefore, a direction of β should be matched by a direction of the hurricane rotation.

The Bretschneider's relationship is used by National Weather Service, NWS (e.g. Phadke et al., 2003; Lin and Chavas, 2012). A similar relationship is suggested by Sobey et al. (1977) to calculate an inflow angle of the hurricane velocity (e.g., Martino et al., 2001; Hu et al., 2012; Lin and Chavas, 2012).

9.10.6 Hurricane wind velocity adjustment for hurricane forward motion

Hurricane velocity calculated by parametric hurricane models is a rotational value of the hurricane velocity for a stationary storm. However, most of the hurricanes have a translation motion. In such cases, hurricane velocity consists of the rotational velocity and translation velocity. Therefore, for a moving storm, after rotational velocity of the hurricane is calculated, hurricane translation velocity (also known as forward velocity or background wind) needs to be added as a vector to rotational velocity (Hu et al., 2012). One of the widely used methods for calculating and adding background wind velocity to hurricane rotational velocity is the one used by SLOSH model as (Jelesnianski et al., 1992; Martino et al., 2001; Cheung et al., 2007; Wei, 2015):

$$U(r) = \frac{R_{mw} \times r}{R_{mw}^2 + r^2} \times V_T \qquad (9.145)$$

where, U is a velocity correction to account for hurricane translation wind velocity at a distance r from a hurricane center, V_T is hurricane translation velocity (also known as forward velocity or background wind), r is radius (distance from a hurricane center), R_{mw} is a radius (distance to a location) of the maximum wind velocity.

After values of U are calculated, then the total velocity for a moving hurricane at each point is obtained by adding hurricane rotational velocity, V, and hurricane translation correction velocity, U, in a vector form as $\vec{V} + \vec{U}$. Note that, both V and U must be at the same height.

Lin and Chavas (2012) suggested another method for considering an effect of the hurricane background velocity on hurricane rotational velocity. They suggested adding a vector with a magnitude of $0.55V_T$ and an inflow angle of 20° (a counter-clockwise rotation of 20° respected to a V_T direction in the Northern hemisphere) to the rotational velocity of the hurricane, vectorially (e.g. Orton et al., 2016).

In addition to previous methods, there is another approach for considering an effect of the hurricane background velocity. This method suggests adding a constant value of V_T as a vector to all rotational velocities of the hurricane (e.g. Coastal Engineering Manual, CEM, 2015).

To use any of these methods, it is important to make sure that during this procedure all considered velocities are at the same height.

9.11 SWAN wave model

As the performance and speed of the computers enhanced significantly during recent years, it becomes more and more appealing for researchers to use computers to simulate phenomena in water bodies numerically. Nowadays, there is a series of commercial and open source software and packages that are intended for wave, tide and storm surge modeling. Some of the most notable ones are listed in the following table (e.g. Thomas and Dwarakish, 2015).

SWAN (Booij et al., 1999) wave model is one of the widely used wave models by researchers. SWAN is the third-generation spectral phase-averaging model, which is developed at Delft University of Technology (e.g. Thomas and Dwarakish, 2015).

In this section, it is described how to set up and run a simple model using SWAN. Most wave models, including SWAN, require sets of input data such as initial and boundary conditions (like bathymetry), water surface elevation, and wind data. Also, wave models require information about simulation settings and procedures such as a numerical scheme, a calculation grid, and physical models to be used in calculations. All these inputs need to be prepared properly before running a model.

Before preparing required inputs, SWAN wave model needs to be installed. To install SWAN, obtain SWAN installing file(s) from its website (swanmodel.sourceforge.net) and follow a provided instruction. On a Windows machine, it is as simple as downloading and running an installer. Both a single core and multiple cores (parallel) versions are available, however, it might be good to start with a single core version.

Table 9.3 Wave numerical models in an alphabetical order

Wave Numerical Model Name	Developer
ADCIRC	ADCIRC, adcirc.org
Delft3D	Deltares, www.deltares.nl
FUNWAVE-TVD	Kirby et al. (1998), fengyanshi.github.io
FVCOM	Chen et al. (2012), fvcom.smast.umassd.edu
IHFOAM	IHCantabria, ihfoam.ihcantabria.com
MIKE	DHI, www.dhigroup.com
PSOM	Mahadevan (2006), github.com/PSOM
ROMS	Shchepetkin and McWilliams, 2005, www.myroms.org
SLOSH	Jelesnianski et al. (1992), www.nhc.noaa.gov/surge/slosh.php
SMS	Aquaveo, www.aquaveo.com
SWAN	Booij et al. (1999), swanmodel.sourceforge.net
SWASH	Zijlema et al. (2011), swash.sourceforge.net
TELEMAC	TELEMAC, www.opentelemac.org
XBeach	Deltares, www.deltares.nl
WAVEWATCH III	Tolman (2014), polar.ncep.noaa.gov/waves/index2.shtml

9.11.1 Preparing a computational grid for SWAN

In this section, it is described how to generate a computational grid for SWAN. In this example, an area of the interest to be modeled in SWAN is the Terrebonne Bay located in the North of Gulf of Mexico. A calculation domain in this example has a longitude range of $-91.5° \leq \lambda \leq -89.5°$ and a latitude range of $28.5° \leq \phi \leq 29.5°$.

To accurately simulate waves in this particular area, a mesh size equal to 0.001° is required. However, since this example is for a demonstration purpose, a larger mesh size equal to 0.01° is chosen for a computational grid. This helps to reduce required calculation time and memory. Note that a chosen mesh size of 0.01° is approximately equal to 1111 m, which is large for this specific area, and can result in incorrect outcomes. Note that, a mesh size can be converted from the radian to meter by using a circle arc equation formula as $Length\ in\ meter = 6.371 \times 10^6 \times Angle\ in\ radian$, where $6.371 \times 10^6\ m$ is the Earth average radius.

Selecting a proper mesh size is highly important in order to get accurate results in any simulation. In general, a smaller mesh leads to a more stable and more accurate simulation. However, by reducing a mesh size, required calculation time and memory can increase dramatically. To find the best mesh size, usually, several simulations with different mesh sizes are run. Then, outcomes of these simulations are evaluated and compared with known or reliable values. The largest mesh size that produces acceptable and accurate results is commonly chosen for the simulation.

To generate a gridded mesh as $(xgrid, ygrid)$, the *meshgrid* command in MATLAB, GNU Octave, or Python's Numpy package can be used, where $xgrid$ and $ygrid$ are 2-dimensional arrays that contain x and y locations of the grid points, respectively. A gridded mesh $(xgrid, ygrid)$ is generated in MATLAB or GNU Octave as:

```
%Generating a gridded mesh
[xgrid,ygrid]=meshgrid(linspace(-91.5,-
89.5,200),linspace(28.5,29.5,100));
```

Here is a Python code for generating a gridded mesh $(xgrid, ygrid)$. Note that you need to install and import NumPy package as np along with Python.

```
#Importing required packages
import numpy as np

#Generating a gridded mesh
xgrid,ygrid=np.meshgrid(np.linspace(-91.5,-
89.5,200),np.linspace(28.5,29.5,100));
```

Generated $xgrid$ and $ygrid$, both, have an array size of 100×200.

After $xgrid$ and $ygrid$ are generated, they need to be re-formatted to match a format required by SWAN. To show how to do that, a sample grid, formatted for SWAN, is demonstrated in the following example. Assume a grid of 4×4 needs to be generated for SWAN. In that case, a 4×4 $xgrid$ would be as:

```
x1 x2 x3 x4
x1 x2 x3 x4
x1 x2 x3 x4
x1 x2 x3 x4
```

Also, 4×4 $ygrid$ would be as:

```
y1 y1 y1 y1
y2 y2 y2 y2
y3 y3 y3 y3
y4 y4 y4 y4
```

Then, a grid array for SWAN would be:

```
x1 x2 x3 x4
x1 x2 x3 x4
x1 x2 x3 x4
x1 x2 x3 x4
y1 y1 y1 y1
y2 y2 y2 y2
y3 y3 y3 y3
y4 y4 y4 y4
```

Similar to this example, SWAN computational grid array should be generated from $xgrid$ and $ygrid$ following an above format. After SWAN grid array is generated from $xgrid$ and $ygrid$, this grid array needs to be saved as ASCII file with ".xy" extension. To save an array as a file, the *dlmwrite* command in MATLAB, and GNU Octave, or the *savetxt* command in Python's Numpy package can be used. In this example, a grid file is saved as "swangrid.xy".

9.11.2 Preparing depth data for SWAN

Bathymetry data is one the inputs required in most wave simulations. Bathymetry data are generally available as a series of scatter data points in the form of (x, y, z), where (x, y) represents a location of the point and z represents an elevation of that point. If a geographic coordinate system is used, then x is presented by a longitude, λ, and y is presented by a latitude, ϕ. In this example, bathymetry data for a calculation domain are obtained from NOAA ETOPO1 global relief model, (Amante and Eakins, 2009), www.ngdc.noaa.gov/mgg/global/global.html. Note that a vertical datum for NOAA's ETOPO1 is at the sea level. A calculation domain has a longitude range of $-91.5° \le \lambda \le -89.5°$ and a latitude range of $28.5° \le \phi \le 29.5°$.

To prepare bathymetry data for SWAN, these scatter (x, y, z) data points should be interpolated to the gridded data points. To interpolate scatter (x, y, z) data on a gridded $(xgrid, ygrid)$ mesh, first, (x, y, z) data are imported. Then, the *griddata* command in MATLAB, GNU Octave, or Python's Scipy package is used to interpolate them on $(xgrid, ygrid)$ mesh. After (x, y, z) data are imported, elevation data are interpolated in MATLAB or GNU Octave as:

```
%Interpolating elevation data, z, on a gridded mesh
(xgrid,ygrid)
zgrid=griddata(x,y,z,xgrid,ygrid);
```

Here is a Python code for interpolating elevation data on a gridded $(xgrid, ygrid)$ mesh. Note that you need to install and import SciPy package as sp along with Python.

```
#Importing required packages
import scipy as sp
from scipy import interpolate

#Interpolating elevation data, z, on a gridded mesh (xgrid,ygrid)
zgrid=sp.interpolate.griddata((x,y),z,(xgrid,ygrid));
```

A generated *zgrid* has an array size of 100 × 200.

After $zgrid$ array is generated, it needs to be converted to depth array. SWAN needs depth data, where values below a water surface are positive, and values above a water surface are negative. For example, if a point has a depth equal to $2\,m$, it means that point is located $2\,m$ below a water surface. However, bathymetry data mostly have an upward vertical axis. It means that positive values in bathymetry data are referred to points that are above a water surface, and negative values are referred to points that are below a water surface. If bathymetry data have an upward vertical axis, then depth values would be obtained as:

$$swandepth = -zgrid$$

After a depth array is obtained, a depth array needs to be saved as ASCII file with ".dep" extension. To save an array as a file, the *dlmwrite* command in MATLAB and GNU Octave, or the *savetxt* command in Python's Numpy package can be used. In this example, a depth file is saved as "swandepth.dep".

9.11.3 Preparing water level data for SWAN

Water level data is required for most wave simulations. By having both bathymetry data and water level data, SWAN can calculate a total water depth. Water level data are available in a vast number of scatter meteorological stations throughout the globe in the form of time series. Note that, it is extremely important that both bathymetry data and water level data have the same vertical datum, otherwise, calculated total water depth would be incorrect. If a vertical datum of the bathymetry data is different from a vertical datum of the water level data, then a vertical datum of one needs to be converted to a vertical datum of the other one. One simple way to do that, is to use NOAA's Vertical Datums Transformation software, VDatum (vdatum.noaa.gov), to find a transformation value (a vertical difference) from one datum to another one.

In SWAN, water level data should have an upward vertical axis. It means that positive values are referred to water level values that are above a vertical datum, and negative values are referred to water level values that are below a vertical datum. Note that, in SWAN, depth data have a downward vertical axis.

If at each time step, a water level is constant (not changing) throughout an area of the study, a single value for a water level is used at each time step. However, if at each time step, a water level is changing throughout an area of the study, then water level data should be interpolated on a gridded $(xgrid, ygrid)$ mesh for each time step. For example, such conditions may occur when a storm surge passes through an area of interest.

The intent of this simulation is to calculate wind waves from Oct 1, 2017 to Oct 31, 2017 in UTC (Coordinated Universal Time). In this example, water level data are obtained from NOAA's Tides and Currents (tidesandcurrents.noaa.gov), at Grand Isle station (Station ID: 8761724) from Oct 1, 2017 to Oct 31, 2017 UTC, in 1 *hr* interval, with the MSL (mean sea level) vertical datum.

In this example, it is assumed that at each time step, a water level is not changing throughout a calculation domain. To prepare water level data for SWAN, a rectangular domain, which should be larger than a calculation domain, is defined. This domain should cover a computational domain entirely. In this example, first, a water level domain is defined as a rectangle with a lower-left corner located at $(-92°, 28°)$ and an upper-right corner located at $(-87°, 33°)$. Then, a value of the water level at each time step is assigned to all four corners of the water level domain.

To show how to do that, a sample water level domain, formatted for SWAN, is demonstrated in the following example. Assume a water level time series has 3 time steps with values of $wl1$, $wl2$, and $wl3$. Then a water level array for SWAN would be:

```
wl1 wl1
wl1 wl1
wl2 wl2
wl2 wl2
wl3 wl3
wl3 wl3
```

Similar to this example, SWAN water level array should be generated following an above format. After SWAN water level array is generated, this water level array needs to be saved as ASCII file with ".wl" extension. To save an array as a file, the *dlmwrite* command in MATLAB and GNU Octave, or the *savetxt* command in Python's Numpy package can be used. In this example, a water level file is saved as "swanwaterlevel.wl".

9.11.4 Preparing wind data for SWAN

Wind data is another input required in most wave simulations. Similar to water level, wind data are available in a vast number of scatter meteorological stations throughout the globe in the form of time series. Generally, wind data can be obtained from the same meteorological station that a water level is obtained from. An important point about wind velocity data is that, wind data should represent 10-min averaged wind velocity at 10 m height above a surface. If wind data obtained from a meteorological station have different settings, they need to be converted to this standard setting.

In this example, wind data are obtained from NOAA's Tides and Currents (tidesandcurrents.noaa.gov), at Grand Isle station (Station ID: 8761724) from Oct 1, 2017 to Oct 31, 2017 UTC, in 1 hr interval. For the station used in this example, i.e. Grand Isle station, a meteorological site is about 2.68 m (8.8 ft) above the MSL (mean sea level), and a wind anemometer is about 6.55 m (21.5 ft) above a meteorological site elevation. In total, a wind anemometer is about 9.23 m (30.3 ft) above a sea level. Therefore, wind velocity data obtained from this station is converted from 9.23 m above a surface to 10 m above a surface by using a power law wind velocity profile as $U_{z=10} = U_{z=9.23}(10/9.23)^{(1/7)}$.

Also, wind data from the Grand Isle station are 2-min averaged. Therefore, wind velocity data need to be converted from 2-min averaged velocity to 10-min averaged velocity. To do that, a wind gust factor suggested by Coastal Engineering Manual, CEM (2015), equal to $G = 0.898$ is used as $U_{10\ min-avg} = G \times U_{2\ min-avg}$.

If at each time step, the wind is constant (not changing) throughout an area of the study, a single value for wind velocity in each of x and y direction is used at each time step. However, if at each time step, the wind is changing throughout an area of the study, then wind velocities in each of x and y direction should be interpolated on a gridded $(xgrid, ygrid)$ mesh for each time step. For example, such conditions may occur when a hurricane passes through an area of interest.

In this example, it is assumed that at each time step, wind velocity is not changing throughout a calculation domain. To prepare wind data for SWAN, a rectangular domain, which should be larger than a calculation domain, is defined. This domain should cover a computational domain entirely. In this example, a wind domain is defined as a rectangle with a lower-left corner located at $(-92°, 28°)$ and an upper-right corner located at $(-87°, 33°)$. Then, values of the wind velocities in x and y directions at each time step are assigned to all four corners of the wind domain.

To show how to do that, a sample wind domain, formatted for SWAN, is demonstrated in the following example. Assume a wind velocity time series has 3 time steps with values of $(U1, V1)$, $(U2, V2)$, and $(U3, V3)$, where U and V represent wind velocities in x and y directions, respectively. Then, a wind array for SWAN would be:

```
U1 U1
U1 U1
V1 V1
V1 V1
U2 U2
U2 U2
V2 V2
V2 V2
U3 U3
U3 U3
V3 V3
V3 V3
```

Similar to this example, SWAN wind array should be generated following an above format. After SWAN wind array is generated, this wind velocity array needs to be saved as ASCII file with ".wnd" extension. To save an array as a file, the *dlmwrite* command in MATLAB and GNU Octave, or the *savetxt* command in Python's Numpy package can be used. In this example, a wind file is saved as "swanwind.wnd".

9.11.5 Preparing a command input file for SWAN

In order to start simulating with SWAN, it is required to pass specific information such as a name and a location of the input data, a numerical scheme, and output results to SWAN. These parameters and information need to be written in an ASCII file with ".swn" extension. In this example, a command input file containing this information is named "swaninput.swn". This file can be edited with any text editor. Note that, in this file, any line starts with the $ sign is considered as a comment line, and is not executed by SWAN. Only some of the SWAN commands are explained here. For a list of all commands, further explanation on how to use each command, and possible changes refer to SWAN user manual (swanmodel.sourceforge.net).

Commands that are explained here are based on the SWAN user manual for SWAN version 41.20.

Before using a SWAN command input file, make sure to review a proper name length described in a SWAN user manual. SWAN has a specific requirement for a maximum character length allowed for names and will return an error if a length requirement does not comply. Also, make sure there is no extra space at the end of each line. In other words, make sure that each line finishes with a character, not space.

The first section in a SWAN command input file contains start-up commands. In this section, PROJect command defines a project name. SET command defines general settings. MODE command defines a stationary status and a number of dimensions. COORDINATES command defines a coordinate system to be used.

```
$Start up---------------------------------------------------------------

$Title
PROJect 'TEST' '001'

$Indicates that the Nautical convention to be considered for wind and wave
directions
SET NAUTical

$Indicates that the run is stationary or nonstationary and 1-D or 2-D
MODE NONSTationary TWODimensional

$Choose a coordinate system between Cartesian and spherical coordinates
COORDINATES SPHErical
```

Next section in a SWAN command input file contains model description commands. In this section, CGRID command defines a computational grid. In this example, a computational grid is set as curvilinear with 199 mesh cells in the x direction and 99 mesh cells in the y direction (Note that a generated computational grid has a size of 100×200). Any values equal to -999 will be ignored. A directional spectrum is set to have 36 grids (10° interval). A power spectral density set to have 32 frequency bins with a frequency range from 0.05 Hz to 1.055 Hz. SWAN has a specific equation for defining these frequency limits. READgrid command defines a name of the file that contains a computational grid. This command is only required if a grid is curvilinear, and should follow the CGRID command immediately.

```
$Model description-----------------------------------------------------

$Defines a computational grid (x-y grid, directional grid, frequency grid)
CGRID CURVilinear 199 99 EXCeption -999 -999 CIRcle 36 0.05 1.055 32
READgrid COORdinates 1.0 'swangrid.xy'
```

Next section in a SWAN command input file contains input grids and data commands. In this section, INPgrid command defines properties of the input grid and its time characteristics. A keyword following the INPgrid command defines the type of the grid. For example, INPgrid BOTtom defines that a grid is for a bottom elevation or depth. In this example, a bottom level grid is curvilinear and is the same as a computational grid. A bottom grid is set as curvilinear with 199 mesh cells in the x direction and 99 mesh cells in the y direction (Note that a generated computational grid has a size of 100×200). Any values equal to -999 will be ignored. If an input grid is regular, then a grid does not need to be imported and can be defined and generated by INPgrid command. Wind and water level grids are regular grids consist of 1 cell with 4 grid points, one at each corner. A 1 cell regular grid is started at a lower-left corner (the South-West corner) with coordinates of $(-92.0°, 28.0°)$ and has a positive cell length of 5° in each direction. Note that this grid should be larger than a computational grid and covers it entirely. For a non-stationary input such as wind and water level, start time and time step interval are required as well. In this example, start time is 2017/10/01, 00:00 hour, stop time is 2017/10/31, 23:00 hour, and a time step interval is 60 minutes, which results in a total of 744 time steps. READinp command defines a name of the input variable and a file that contains that variable. This command should follow INPgrid command immediately.

```
$Input grids and data-------------------------------------------------

$Defines an input grid of the bottom level
INPgrid BOTtom CURVilinear 0 0 199 99 EXCeption -999
READinp BOTtom 1.0 'swandepth.dep' 1
```

```
$Defines an input grid of the water level (regular rectangular grid with 4 grid
points)
INPgrid WLEVel REGular -92.0 28.0 0 1 1 5 5 NONSTATionary 20171001.000000
60 MIn 20171031.230000
READinp WLEVel 1.0 'swanwaterlevel.wl'

$Defines an input grid of the wind velocity (regular rectangular grid with 4 grid
points)
INPgrid WInd REGular -92.0 28.0 0 1 1 5 5 NONSTATionary 20171001.000000 60
MIn 20171031.230000
READinp WInd 1.0 'swanwind.wnd'
```

Next section in a SWAN command input file contains boundary and initial conditions commands. This section is not used in this example, however, as a sample BOUnd SHAPespec and BOUndspec commands are explained (Note they are written as comment lines). BOUnd SHAPespec defines a shape of the power spectral density at a boundary of the computational grid for cases that a parametric spectrum is used as an input. BOUndspec defines a location and parameters of the parametric power spectrum. Also, BOUndnest1 command can be used to define nested SWAN run.

```
$Boundary and initial conditions------------------------------------------

$Defines a spectrum shape at a boundary of the computational grid if a
parametric spectrum is used as an input
$BOUnd SHAPespec JONswap 3.3 PEAK DSPR DEGRees

$Defines a location and parameters of the parametric spectrum at a boundary
$BOUndspec SEGMent XY -91.5 28.5 -89.5 28.5 CONstant PAR 0.5 5 180 5
```

Next section in a SWAN command input file contains physics commands. In this section, GEN3 command set the SWAN to run in the third-generation mode. WCAPping defines options related to whitecapping. BREaking defines options for depth-induced wave breaking in shallow water. FRICtion activates a bottom friction option. TRIad activates a triad wave-wave interactions option.

```
$Physics----------------------------------------------------------------

$Indicates that SWAN will be run in the third-generation mode
GEN3 KOMen AGROW

$Defines whitecapping options
WCAPping KOMen

$Defines options for depth-induced wave breaking in shallow water
BREaking CONstant

$Activates bottom friction
FRICtion JONswap CONstant 0.019

$Activates triad wave-wave interactions
TRIad
```

Next section in a SWAN command input file contains numerics commands. In this section, PROP command defines a numerical scheme. NUMeric command defines numerical properties. In this example, it is set to stop a calculation at each time step after 25 iterations, if non-stationary criteria to terminate an iterative process are not met in the first 25 iterations.

```
$Numerics---------------------------------------------------------------

$Defines a numerical scheme
PROP BSBT

$Defines numerical properties
NUMeric STOPC NONSTAT 25
```

Next section in a SWAN command input file contains output locations commands. In this section, FRAme command defines output locations as a regular grid. In this example, frame is set as a regular grid with a name 'TBay'. A lower-left corner (the South-West corner) of the frame has coordinates of $(-91.0°, 29.0°)$. A frame is extended 1° in the x direction and 0.5° in the y direction, with 20 cells in the x direction and 20 cells in the y direction. POINts command defines an output location as a single point. In this example, it defines a point named 'TBayCntr' at coordinates of $(-90.5°, 29.18°)$.

```
$Output locations-------------------------------------------------------

$Defines output locations as a regular grid
FRAme 'TBay' -91.0 29.0 0 1.0 0.5 20 20

$Defines an output location as a single point (coordinates of one output
location)
POINts 'TBayCntr' -90.5 29.18
```

Next section in a SWAN command input file contains write output data commands. In this section, OUTPut command sets a format of the block, table, and spectrum for writing spatial distributions. BLOck command indicates that spatial distributions of the variables should be written to a file. In this example, H_{m0} on a computational grid is written as a text file named 'Hm0compgrid.txt', and T_p on a computational grid is written as a Matlab MAT file named 'Tpcompgrid.mat', starting on 2017/10/01 with a 60-minutes interval. Also, gridded H_{m0} and T_p are outputted on a frame named 'TBay', and are written as Matlab MAT files named 'Hm0grid.mat' and 'Tpgrid.mat', respectively. The 'COMPGRID' keyword indicates that outputs are reported on a computational grid. TABle command indicates that variables for every single location to be written in a tabular format to a file. In this example, outputs for a location named 'TBayCntr' are written in an ASCII file named 'TBayCntr.tab', starting on 2017/10/01 with a 60-minutes interval. Note that, these output names and formats are used as examples, and any other names or supported file formats can be used.

```
$Write output data------------------------------------------------------

$Sets a format of the block for writing spatial distributions (only for FRAMEs and
GROUPs)
OUTPut OPTIons BLOck 4

$Sets a format of the table for writing outputs at output location(s)
OUTPut OPTIons TABle 11

$Indicates spatial distributions of the variables on a computational grid to be
written to a file (ASCII)
BLOck 'COMPGRID' NOHEADer 'xcompgrid.txt' LAYout 3 XP
```

```
BLOck 'COMPGRID' NOHEADer 'ycompgrid.txt' LAYout 3 YP
BLOck 'COMPGRID' NOHEADer 'Hm0compgrid.txt' LAYout 3 HSign OUTput
20171001.000000 60 MIn
$BLOck 'COMPGRID' NOHEADer 'Tpcompgrid.mat' LAYout 3 TPS OUTput
20171001.000000 60 MIn

$Indicates spatial distributions of the variables on a defined grid to be written
to a file
BLOck 'TBay' NOHEADer 'Hm0grid.mat' LAYout 3 HSign OUTput
20171001.000000 60 MIn
BLOck 'TBay' NOHEADer 'Tpgrid.mat' LAYout 3 TPS OUTput 20171001.000000
60 MIn

$Indicates variables for every single location to be written as a table to a file
TABle 'TBayCntr' HEADer 'TBayCntr.tab' TIME XP YP DEPth WIND HSign TPS RTP
DIR PDIR LWAVP OUTput 20171001.000000 60 MIn
```

The last section in a SWAN command input file contains lock up commands. In this section, COMPute command tells SWAN to start a simulation. In this example, SWAN starts a non-stationary simulation from 2017/10/01, 00:00 hour to 2017/10/31, 23:00 hour with a 60-minutes interval, for a total of 744 time steps. Stop command indicates the end of the command input file.

```
$Lock up-----------------------------------------------------------------

$Instructs SWAN to start a simulation
COMPute NONSTat 20171001.000000 60 MIn 20171031.230000

$Indicates the end of the command input file
Stop
```

9.11.6 Start SWAN simulation

After all input files are prepared, they all have to be placed in the same folder. In this example, all required files are placed in a folder at "C:\swan run files". Required files in this example are: a grid file "swangrid.xy", a depth file "swandepth.dep", a water level file "swanwaterlevel.wl", a wind file "swanwind.wnd", and a SWAN command input file "swaninput.swn".

After all input files are prepared and placed in the same folder, a SWAN command prompt needs to be started. To do that, on a Windows machine, first, click on Start, then go to Swan under Programs (Its location might differ depending on the Windows version), locate a file named "SWAN" and click on that. Alternatively, by using the Windows File Explorer, a SWAN command prompt can be started by going a folder that contains SWAN file and double click on that. A folder contains a SWAN file on the Windows 10 is similar to C:\ProgramData\Microsoft\Windows\Start Menu\Programs\Swan.

After a SWAN command prompt is started, use CD command to change a current directory to a directory that all SWAN input files are stored. In this example, it would be:

```
C:>cd C:\swan run files
```

Next, to start a SWAN simulation, in a SWAN command prompt, type swanrun command followed by a SWAN command input file name without ".swn" extension. In this example, it would be:

```
C:\swan run files>swanrun swaninput
```

A simulation will start and continue based on defined criteria in a SWAN command input file. SWAN will save requested outputs in the same folder where all SWAN input files are placed.

9.12 Goodness of fit

In many studies, it is required to evaluate the accuracy of the dataset, such as a model result, by comparing it with a reliable dataset. The accuracy of the dataset or model usually is evaluated by assessing goodness of fit. Some of the widely used measures for goodness of fit are: a relative error, RE, a root-mean-square error, $RMSE$, a scatter index, SI, the Nash–Sutcliffe efficiency coefficient, NSE, the Pearson's correlation coefficient, r, a coefficient of determination, R^2, and a normalized mean bias, NMB.

Considering X as a reliable dataset with true (exact or expected) values such as theoretical values and Y as a dataset that is required to be evaluated such as model results, then the goodness of fit parameters for a dataset which contains N paired samples of (X, Y) are (e.g. Moriasi et al., 2007; Harmel et al., 2010; Bennett et al., 2013; Karimpour and Chen, 2016; Karimpour et al., 2017):

$$RE = \frac{Y-X}{X} = \frac{Y}{X} - 1 \tag{9.146}$$

$$RMSE = \sqrt{\frac{\sum(Y_i-X_i)^2}{N}} \tag{9.147}$$

$$SI = \frac{RMSE}{\bar{X}} \tag{9.148}$$

$$NSE = 1 - \frac{\sum(Y_i-X_i)^2}{\sum(X_i-\bar{X})^2} \tag{9.149}$$

$$r = \frac{\sum(X_i-\bar{X})(Y_i-\bar{Y})}{\sqrt{\sum(X_i-\bar{X})^2 \sum(Y_i-\bar{Y})^2}} \tag{9.150}$$

$$R^2 = r^2 = \left(\frac{\sum(X_i-\bar{X})(Y_i-\bar{Y})}{\sqrt{\sum(X_i-\bar{X})^2 \sum(Y_i-\bar{Y})^2}}\right)^2 \tag{9.151}$$

$$NMB = \frac{\bar{Y}-\bar{X}}{\bar{X}} \tag{9.152}$$

where, X is a dataset with true (exact or expected) values, Y is a dataset that is required to be evaluated, RE is a relative error, $RMSE$ is a root-mean-square error, SI is a scatter index, NSE is the Nash–Sutcliffe efficiency coefficient, r is the Pearson's correlation coefficient, R^2 is a coefficient of determination, NMB is a normalized mean bias, $\bar{X}$ is a mean value of X dataset, and $\bar{Y}$ is a mean value of Y dataset.

A relative error, RE, compares an absolute error, i.e. $(Y - X)$, against an exact value, i.e. X, and shows uncertainty in Y dataset compared to exact values.

A root-mean-square error, $RMSE$, shows differences between exact variables in X dataset and variables in Y dataset.

A scatter index, SI, which is equal to $RMSE$ normalized by a mean of the true values, i.e. $\bar{X}$, shows dispersion respect to a 45° line, i.e. a line with a slope of 1:1.

The Nash–Sutcliffe efficiency coefficient, NSE, shows how well X and Y datasets fit a line with a slope of 1:1. The NSE has a ranges of $-\infty \leq NSE \leq 1$, where $NSE = 1$ indicates a perfect match between X and Y datasets, $NSE = 0$ indicates that predicted values in Y dataset are as accurate as a mean of X dataset, and $NSE < 0$ indicates that a mean of X values is a better predictor than the estimated ones in Y dataset.

The Pearson's correlation coefficient, r, shows a degree of the linear correlation between X and Y values. The Pearson's correlation coefficient has a range of $-1 \leq r \leq 1$, where $r = 1$ indicates a complete positive correlation, $r = 0$ indicates no correlation, and $r = -1$ indicates a complete negative correlation.

A coefficient of determination, R^2, shows how well a model, i.e. Y dataset, fits the data, i.e. X dataset. A coefficient of determination has a range of $0 \leq R \leq 1$, where $R^2 = 1$ indicates a model, i.e. Y dataset, perfectly fits the data, i.e. X dataset, and $R^2 = 0$ indicates that a model, i.e. Y dataset, does not fit the data, i.e. X dataset, at all.

A normalized mean bias, NMB, shows an average tendency of Y dataset compared to X dataset. For positive X and Y values, $NMB = 0$ indicates identical means for X and Y values, i.e. $\bar{X} = \bar{Y}$, a positive NMB, i.e. $NMB > 0$, indicates a model, i.e. Y dataset, has an overestimation bias, and a negative NMB, i.e. $NMB < 0$, indicates a model, i.e. Y dataset, has an underestimation bias (e.g. Moriasi et al., 2007; Harmel et al., 2010; Bennett et al., 2013; Karimpour and Chen, 2016; Karimpour et al., 2017).

Notation

Chapter 1

a	amplitude
a_x	water particles accelerations in x direction
a_z	water particles accelerations in z direction
C	wave celerity or wave phase velocity
C_0	deep-water wave celerity
C_g	group wave velocity
E	mean value of the total energy density per unit horizontal area of the water surface
E_K	mean value of the kinetic energy density per unit horizontal area of the water surface
E_P	mean value of the potential energy density per unit horizontal area of the water surface
f	wave frequency
g	gravitational acceleration
h_0	deep-water depth
h_b	water depth at a breaking location
H	wave height
H_0	deep-water wave height
H_b	wave height at a breaking location
k	wavenumber
k_0	deep-water wavenumber
K_p	pressure response factor (dynamic pressure to a water surface elevation conversion factor)
L	wavelength
L_0	deep-water wavelength
L_b	wavelength at the breaking location
P	static pressure
P_0	total pressure
P_w	wave power
q	dynamic pressure
S_{XX}	radiation stress in x direction
S_{YY}	radiation stress in y direction
t	time
T	wave period
u	water particles velocities in x direction
u_m	maximum near bed orbital velocity

$\bar{u}$	Stokes drift
U	mean water velocity
w	water particles velocities in z direction
x	distance in a horizontal direction
z	upward vertical axis with zero at a water surface
ζ	displacements of the water particles in x (horizontal) direction
η	water surface elevation or excursion
ξ	displacements of the water particles in z (vertical) direction
ρ	water density
σ	intrinsic angular frequency
ϕ	random phase angle or phase shift
ω	angular frequency
ω_0	deep-water angular frequency
h/L	relative water depth
H/h	wave height to water depth ratio
H/L	wave steepness
$(kx - \omega t)$	wave phase

Chapter 2

f	frequency
f_s	sampling frequency
h	local water depth
t	time
t_{bl}	data block duration
t_{br}	burst duration
Δt_{bl}	interval between measured blocks of data
Δt_s	time interval between two sequential data points
η	water surface elevation or excursion

Chapter 3

a	amplitude
f	frequency
f_{pdf}	probability density function
$f_{pdf}(H)$	probability density function of the wave height
H	wave height

$H_{1/3}$	a mean value of the highest one-third of the wave heights
H_e	given wave height
$H_{e(n/N)}$	given wave height ranked n in a wave record with the total of N waves
H_{max}	largest wave height in a wave record
$H_{n/N}$	average wave height of the top n waves
H_{rms}	root mean square (rms) wave height
H_s	significant wave height
H_z	zero-crossing mean wave height
$\overline{H}$	mean wave height
k	wavenumber
n	rank of H_e in a wave record, sorted from the largest ($n = 1$) to the smallest ($n = N$)
N	the total number of waves in a dataset
$P(H \geq H_e)$	probability of the wave height, H, being equal or larger than given wave height, H_e
$P(H \leq H_e)$	probability of the wave height, H, being equal or smaller than given wave height, H_e
$RMSE$	root-mean-square error
t	time
T	wave period
T_s	significant wave period
T_z	zero-crossing mean wave period
x	distance in a horizontal direction
η	water surface elevation
$\eta(t)$	water surface elevation time series
η_c	wave crest height
η_{rms}	root mean square (rms) of the water surface elevation
η_t	wave trough height
σ	variance of the data
ϕ	phase
ω	wave angular frequency

Chapter 4

a	amplitude
a_n	amplitude (Fourier series coefficient)
b_n	amplitude (Fourier series coefficient)
c_n	amplitude (Fourier series coefficient)
d_n	amplitude (Fourier series coefficient)
E	wave energy

E_K	wave kinetic energy
E_P	wave potential energy
f	frequency
f_p	peak wave frequency
f_{max}	high-cutoff frequency (low-pass filter)
f_{min}	low-cutoff frequency (high-pass filter)
f_{Ny}	Nyquist frequency
f_s	sampling frequency
$F(\,)$	continuous Fourier transform
$F[\,]$	discrete Fourier transform
$\mathcal{F}$	Fourier transform operator
$\mathcal{F}^{-1}$	inverse Fourier transform operator
g	gravitational acceleration
H	wave height
H_{m0}	zero-moment wave height
H_n	wave height equal to $2d_n$
i	imaginary unit equal to $\sqrt{-1}$
i	counter
k	wavenumber
m	counter
m_0	zero-moment of the power spectrum
m_n	the nth moment of the power spectrum
n	counter
N	the total number of points or samples
$S_{\eta\eta}$	one-sided water surface elevation power spectral density
$S_{\eta\eta-2S}$	two-sided water surface elevation power spectral density
$S_{\eta\eta-Smoothed}$	smoothed one-sided water surface elevation power spectral density
t	time
T	wave period
T_{m01}	mean wave period
T_{m02}	mean zero-crossing period
T_p	peak wave period
x	distance
Δf	frequency interval
Δt	time interval
η	water surface elevation
$\eta(t)$	time series of the water surface elevation
ρ	water density
σ_η	standard deviation of the surface elevation
ϕ	phase angle or phase shift
ω	angular frequency

Chapter 5

a	amplitude
bw	filter bandwidth
dB	Decibel
f	frequency
f_{cL}	low-cutoff frequency (high-pass filter)
f_{cH}	high-cutoff frequency (low-pass filter)
f_{Ny}	Nyquist frequency
f_p	peak wave frequency
f_s	sampling frequency
F	function in a convolution operator
F	dataset
$\mathcal{F}$	Fourier transform
$F_{Filtered}$	filtered values of F
$h(\,)$	digital filter
$h_{BandPassIR}$	impulse response of the band-pass windowed *sinc* filter
$h_{BandRejectIR}$	impulse response of the band-reject windowed *sinc* filter
$h_{Frequency\ Response}$	frequency response of the filter
$h_{HighPassFR}$	frequency response of the high-pass windowed *sinc* filter
$\lvert h_{HighPassFR}(f)\rvert$	magnitude or gain of the high-pass frequency response
$h_{HighPassIR}$	impulse response of the high-pass windowed *sinc* filter
$h_{Imluse\ Response}$	impulse response of the filter
$h_{LowPassFR}$	frequency response of the low-pass windowed *sinc* filter
$\lvert h_{LowPassFR}(f)\rvert$	magnitude or gain of the low-pass frequency response
$h_{LowPassIR}$	impulse response of the low-pass windowed *sinc* filter
G	function in a convolution operator
n	counter
N	the total number of points or samples
n_w	element counter in a window function
N_w	the total number of points in a window function or in a filter
P	power
$S_{\eta\eta}$	one-sided water surface elevation power spectral density
$S_{\eta\eta-Smoothed}$	smoothed one-sided water surface elevation power spectral density
t	time
t	variable in a convolution operator
U	compensating factor for energy loss due to use of window function in a periodogram

W	window function
W_N	normalized window function
Δf	frequency interval
Δt	time interval
η	water surface elevation
$\eta(t)$	time series of the water surface elevation
τ	variable in a convolution operator
$*$	convolution operation

Chapter 6

a	amplitudes
d_s	pressure measurement distance (pressure sensor location) from a seabed
f	frequency
f_{cH}	high-cutoff frequency (low-pass filter)
f_{cL}	low-cutoff frequency (high-pass filter)
f_m	mean wave frequency
$f_{maxpcorr}$	frequency associated with K_{pmin}
$f_{maxpcorr-Linear}$	the $f_{maxpcorr}$ estimated by the linear wave theory
$f_{MinPoint}$	frequency associated with the minimum value of spectrum at its tail
f_{mPM}	mean wave frequency of the Pierson-Moskowitz spectrum
$f_{NoiseFloor}$	frequency that a noise floor starts at a tail of spectrum
f_p	peak wave frequency
f_s	sampling frequency
f_{tail}	frequency where a spectrum tail after that is replaced with an empirical spectrum tail
g	gravitational acceleration
h	water depth
h_s	sensor depth
k	wavenumber
$k_{max-Linear}$	wavenumber of the smallest linear wave that can be detected by a pressure sensor

K_p	pressure response factor (dynamic pressure to a water surface elevation conversion factor)
K_{pmin}	minimum value of K_p to prevent unacceptable amplification of the wave energy
$K_{pmin-Linear}$	the K_{pmin} estimated by the linear wave theory
L	wavelength
$L_{min-Linear}$	wavelength of the smallest linear wave that can be detected by a pressure sensor
m_0	zero-moment of the wave energy spectrum
m_n	the nth moment of the wave energy spectrum
n	spectrum tail power coefficient
N_{Cor}	correction factor
P	dynamic pressure
P_0	hydrostatic pressure
P_T	total pressure
S_{PP}	dynamic pressure power spectral density
$S_{\eta\eta}$	water surface elevation power spectral density
T	wave period
T_{m01}	mean wave period
z	upward vertical axis with zero at a water surface
Δf_{tr}	frequency range before and after $f_{maxpcorr}$ for K_{pmin} transition
η	water surface elevation
$\eta_{NotCorrected}$	water surface elevation obtained from pressure data without pressure loss correction
ρ	water density
ϕ	phase
Φ	transformation function from JONSWAP spectrum into TMA spectrum
ω	angular frequency

Chapter 7

a	amplitude
d_s	velocity measurement distance (velocity probe location) from a seabed
f	frequency

f_{cH}	high-cutoff frequency (low-pass filter)
f_{cL}	low-cutoff frequency (high-pass filter)
f_m	mean wave frequency
$f_{maxuvcorr}$	frequency associated with K_{uvmin}
$f_{maxuvcorr-Linear}$	the $f_{maxuvcorr}$ estimated by the linear wave theory
$f_{MinPoint}$	frequency associated with the minimum value of spectrum at its tail
f_{mPM}	mean wave frequency of the Pierson-Moskowitz spectrum
$f_{NoiseFloor}$	frequency that a noise floor starts at the tail of spectrum
f_p	peak wave frequency
f_s	sampling frequency
f_{tail}	frequency where a spectrum tail after that is replaced with an empirical spectrum tail
h	water depth
h_s	sensor depth
H	wave height
k	wavenumber
$k_{max-Linear}$	wavenumber of the smallest linear wave that can be detected by a velocity probe
K_{uv}	orbital velocity conversion factor (orbital velocity to a water surface elevation conversion factor)
K_{uvmin}	minimum value of K_{uv} to prevent unacceptable amplification of the wave energy
$K_{uvmin-Linear}$	the K_{uvmin} estimated by linear wave theory
L	wavelength
$L_{min-Linear}$	wavelength of the smallest linear wave that can be detected by a velocity probe
m_0	zero-moment of the wave energy spectrum
m_n	the nth moment of the wave energy spectrum
n	spectrum tail power coefficient
S_{uu}	horizontal orbital velocity power spectral density in x direction
S_{uv}	combined horizontal orbital velocities power spectral density
S_{vv}	horizontal orbital velocity power spectral density in y direction

$S_{\eta\eta}$	water surface elevation power spectral density
t	time
T	wave period
T_{m01}	mean wave period
u	horizontal orbital velocity in x direction
U	de-trended horizontal velocity (horizontal component of the orbital velocity)
U_c	mean current horizontal velocity
U_T	total horizontal velocity
v	horizontal orbital velocity in y direction
x	distance
z	upward vertical axis with zero at a water surface
Δf_{tr}	frequency range before and after $f_{maxuvcorr}$ for K_{uvmin} transition
η	water surface elevation
ϕ	phase
Φ	transformation function from JONSWAP spectrum into TMA spectrum
ω	angular frequency

Chapter 8

a	wave amplitude
a_1	Fourier coefficient
a_2	Fourier coefficient
b_1	Fourier coefficient
b_2	Fourier coefficient
C	wave celerity or wave phase speed
C_p	peak wave celerity
d_p	pressure sensor location distance from a seabed
d_{uv}	velocity probe location distance from a seabed
df	frequency interval
dt	time interval
$D(f,\theta)$	directional spreading function
E_w	wave energy
f	frequency
f_l	lower limit of the wave spectrum
f_{m1}	frequency associated with a maximum value of I_1

f_p	peak wave frequency
$f_{pdf}(H)$	probability density function of the wave height
$f_{pdf}(T)$	probability density function of the wave period
$f_{pdf}(\tau)$	probability density function of the dimensionless wave period
$f_{pdf}(R,\tau)$	joint probability density function for the dimensionless wave amplitude and dimensionless wave period
f_s	sampling frequency
f_{sep}	frequency that separates swell waves from wind sea
f_u	upper limit of the wave spectrum
f_x	frequency associated with a maximum value of $\xi(f)$
$\mathcal{F}$	Fourier transform
g	gravitational acceleration
h	water depth
h_p	pressure sensor depth
h_{uv}	velocity probe depth
H	wave height
H_{m0}	zero-moment wave height
H_{rms}	root mean square (rms) wave height
H_s	significant wave height
i	imaginary unit equal to $\sqrt{-1}$
k	wavenumber
k_0	deep-water wavenumber
k_{est}	estimated wavenumber
k_{exa}	exact wavenumber
k_p	peak wavenumber
K_p	pressure response factor (dynamic pressure to a water surface elevation conversion factor)
K_{uv}	orbital velocity conversion factor (horizontal orbital velocity to a water surface elevation conversion factor)
L	wavelength
L_v	probability density function normalization factor
m_0	zero-moment of the water surface elevation power spectral density
m_n	the nth moment of the water surface elevation power spectral density
n	counter
N	the total number of data points
P	water dynamic pressure
$r(f)$	absolute value of Fourier transform
s	spreading parameter (directional spreading of the wave energy around θ_m)

s_p	peak wave spreading parameter
S_{PP}	water dynamic pressure power spectral density
S_{Pu}	cross power spectral density of the water dynamic pressure and water particle horizontal orbital velocity in x direction
S_{Pv}	cross power spectral density of the water dynamic pressure and water particle horizontal orbital velocity in y direction
S_{uu}	water particle horizontal orbital velocity power spectral density in x direction
S_{uv}	cross power spectral density of the water particle horizontal orbital velocity in x and water particle horizontal orbital velocity in y direction
S_{vv}	water particle horizontal orbital velocity power spectral density in y direction
$S_{\eta u}$	cross power spectral density of the water surface elevation and water particle horizontal orbital velocity in x direction
$S_{\eta v}$	cross power spectral density of the water surface elevation and water particle horizontal orbital velocity in y direction
$S_{\eta\eta}$	water surface elevation power spectral density
$S_{\eta\eta}(f,\theta)$	directional water surface elevation power spectral density
R	dimensionless wave amplitude
t	time
T	wave period
T_{m02}	mean zero-crossing wave period
T_p	peak wave period
T_s	significant wave period
T_z	zero-crossing mean wave period
$\bar{T}$	mean wave period
U	mean flow velocity
U_{10}	10-minute averaged wind velocity at 10 m height above a surface
U_w	wind velocity
x	horizontal direction
y	horizontal direction
z	upward vertical axis with zero at a water surface
β	spreading parameter
Γ	Gamma function
ε	relative error
η	water surface elevation
η_{rms}	root mean square of the water surface elevation
θ	wave direction

θ_1	mean wave direction as a function of frequency
θ_2	principal wave direction as a function of frequency
θ_m	mean wave direction
ν	spectral bandwidth
$\xi(f)$	wave steepness function
ρ	water density
τ	dimensionless wave period
ϕ	wave phase
ω	wave angular frequency

Chapter 9

Az	azimuth (bearing)
C_p	peak wave celerity
$dist$	distance between two points
$\hat{E}$	dimensionless wave energy
$\hat{E}_{Asympt}$	dimensionless asymptotic limit for wave energy
$\hat{E}_{FullyDev}$	dimensionless value for a fully developed wave energy
f	frequency
$\widehat{f_p}_{Asympt}$	dimensionless asymptotic limit for a peak wave frequency
$\widehat{f_p}_{FullyDev}$	dimensionless value for a fully developed peak wave frequency
f_s	sampling frequency
f_p	peak wave frequency (frequency associated with a maximum value of the $S_{\eta\eta}$)
$\widehat{f_p}$	dimensionless peak wave frequency
F	wind fetch
F_{eff}	effective wind fetch
F_{eqv}	equivalent wind fetch
F_{km}	wind fetch in *kilometer*
$\hat{F}$	dimensionless wind fetch
g	gravitational acceleration
g_t	peak factor
G	wind gust factor
h	water depth
h	spatial water depth along a wind fetch axis
$\bar{h}$	mean water depth
$\hat{h}$	dimensionless water depth
H_{m0}	zero-moment wave height
H_s	significant wave height

$\overline{H}$	mean wave height
I_u	wind longitudinal turbulence intensity
k	wavenumber
k_p	peak wavenumber (a wavenumber associated with a peak wave frequency)
$\hat{k}_p$	dimensionless peak wavenumber
K	conversion factor
L_p	peak wavelength (a wavelength associated with a peak wave frequency)
m_0	zero moment of the water surface elevation power spectral density (wave spectrum)
m_{01}	the first moment of the water surface elevation power spectral density (wave spectrum)
m_{02}	the second moment of the water surface elevation power spectral density (wave spectrum)
m_n	the nth moment of the water surface elevation power spectral density (wave spectrum)
N	number of data points
NMB	normalized mean bias
NSE	Nash–Sutcliffe efficiency coefficient
P	air pressure at a surface level at a distance r from a hurricane center
P_c	air pressure at a hurricane center (central pressure)
P_n	ambient pressure (external pressure)
r	radius
r	Pearson's correlation coefficient
R	earth radius
R^2	coefficient of determination
R_{mw}	radius (distance) to a location of the maximum wind velocity
RE	relative error
$RMSE$	root-mean-square error
$S_{\eta\eta}$	water surface elevation power spectral density
SI	scatter index
t	time
t_{min}	minimum duration required for a wave to travel a wind fetch
t_{min-Hr}	minimum duration required for a wave to travel a wind fetch in *hour*
t_{st}	sustained wind duration
$\hat{t}$	dimensionless time

$\hat{t}_{min}$	non-dimensional minimum duration required for a wave to travel a wind fetch
T	wave period
T_{m01}	mean wave period
T_{m02}	mean zero-crossing period
T_p	peak wave period
T_s	significant wave period
$\bar{T}$	mean wave period
u_*	friction velocity (shear velocity)
U	wind velocity
$\bar{U}$	mean wind velocity
U_{10}	10-minute averaged wind velocity at 10 m height above a surface
$U_{10-1min}$	1-minute averaged wind velocity at 10 m height above a surface
$U_{19.5}$	wind velocity measured at 19.5 m above a surface
U_{120}	wind velocity averaged over a duration of 120 seconds
U_{600}	wind velocity averaged over a duration of 600 seconds
U_{3600}	wind velocity averaged over a duration of 3600 seconds
U_A	adjusted wind velocity
U_i	the ith data point in a wind velocity dataset
U_t	wind velocity averaged over a duration of t seconds
V_g	hurricane gradient wind velocity at a distance r from a hurricane center
V_{g-1min}	1-minute averaged (1-minute sustained) gradient wind velocity
V_{gmax}	hurricane maximum gradient wind velocity
V_T	hurricane translation velocity
x	longitudinal axis along a wind fetch
x	horizontal direction (axis)
X	dataset with true (exact or expected) values
$\bar{X}$	mean value of X
y	horizontal direction (axis)
Y	dataset that is required to be evaluated
$\bar{Y}$	mean value of Y
z	upward vertical axis (height from a surface)
z_0	surface roughness
z_d	zero-plane displacement distance
z_{ref}	reference height
α	empirical constant
α	Phillip's constant
α	power law exponent

α_{Ch}	empirical constant (Charnock parameter)
β	empirical constant
β	inflow angle of the hurricane velocity
γ	JONSWAP's peak enhancement parameter
ΔP	pressure drop from pressure outside a hurricane to pressure at a center of the hurricane
$\Delta\sigma$	central angle between points
θ	direction
θ_i	the ith data point in a wind direction dataset
$\bar{\theta}$	mean direction
κ	von Karman's constant (≈ 0.41)
λ	longitude
ρ	density
ρ_a	air density
σ	JONSWAP's spectral width parameter
σ_u	standard deviation of the longitudinal wind velocity
τ	surface shear stress
ϕ	latitude
Φ	TMA spectrum transformation function

References

Chapter 1

Sorensen, R. M. (2005). Basic coastal engineering (Vol. 10). Springer Science & Business Media.

Chapter 2

Jones, N. L., & Monismith, S. G. (2007). Measuring short-period wind waves in a tidally forced environment with a subsurface pressure gauge. Limnology and Oceanography: Methods, 5(10), 317-327.

Chapter 3

Dean, R. G., & Dalrymple, R. A. (1991). Water wave mechanics for engineers and scientists (Vol. 2). World Scientific Publishing Company.

Holthuijsen, L. H. (2007). Waves in oceanic and coastal waters. Cambridge University Press.

Longuet-Higgins, M.S. (1952). On the statistical distribution of the heights of sea waves. Journal of Marine Research, 11: 245-266.

McCormick, M. E. (2009). Ocean engineering mechanics: with applications. Cambridge University Press.

Young, I. R. (1999). Wind generated ocean waves (Vol. 2). Elsevier.

Chapter 4

Chakrabarti, S. K. (1987). Hydrodynamics of offshore structures. WIT press.

Dyke, P. P. (2014). An introduction to Laplace transforms and Fourier series. New York, NY, USA: Springer.

Earle, M. D. (1996). Nondirectional and directional wave data analysis procedures. NDBC technical Document, 96-01.

Eaton J. W., Bateman D., Hauberg S., and Wehbring R. (2018). GNU Octave version 4.2.2 manual: a high-level interactive language for numerical computations.

Hasselmann, K.; Barnett, T. P.; Bouws, E.; Carlson, H.; Cartwright, D. E.; Enke, K.; Ewing, J. A.; Gienapp, H.; Hasselmann, D. E.; Kruseman, P.; Meerbrug, A.; Muller, P.; Olbers, D. J.; Richter, K.; Sell, W., and Walden, H., (1973). Measurements of wind-wave growth and swell decay during the Joint North Sea Wave Project (JONSWAP). Deutsche Hydrographische Zeitschrift A80(12), 95p.

Holthuijsen, L. H. (2007). Waves in oceanic and coastal waters. Cambridge University Press.

Hunter, J. D. (2007). Matplotlib: A 2D graphics environment. Computing in science and engineering, 9(3), 90-95.

Jones E, Oliphant E, Peterson P, et al. (2001). SciPy: Open Source Scientific Tools for Python.

Kawauchi H., Ueno K., Sato M., Fanz C., and Hiraoka R. (2009). Comparative Study of Ocean Wave Height Estimation with Power Spectral Density Function by Three Methods. Japan Society for Mathematical and Physical Fisheries Science, Vol 7, Pages 70-89.

McKinney, W. (2010). Data structures for statistical computing in python. In Proceedings of the 9th Python in Science Conference (Vol. 445, pp. 51-56).

Mobley, C. D., (2014). Modeling Sea Surfaces: A Tutorial on Fourier Transform Methods. Sequoia Scientific, Inc. Bellevue, WA 98005. 71 pages.

O'neil, P. V. (2011). Advanced engineering mathematics. Cengage learning.

Reeve, D., Chadwick, A., and Fleming, C. (2012). Coastal engineering: Process, theory, and design practice, 2nd Ed., Spon, New York.

Shima, H., & Nakayama, T. (2010). Higher mathematics for physics and engineering. Springer Science & Business Media.

Smith, J. O. (2007). Mathematics of the discrete Fourier transform (DFT): with audio applicaitons. Julius Smith.

Van Der Walt, S., Colbert, S. C., & Varoquaux, G. (2011). The NumPy array: a structure for efficient numerical computation. Computing in Science & Engineering, 13(2), 22-30.

Young, I.R., (1995). The determination of confidence limits associated with estimates of the spectral peak frequency. Ocean Engineering, 22, 669-686.

Chapter 5

Bartlett, M. S. (1948). Smoothing periodograms from time series with continuous spectra. Nature, 161(4096), 686-687.

Bartlett, M. S. (1950). Periodogram analysis and continuous spectra. Biometrika, 37(1/2), 1-16.

Djuric, P. M., Kay, S. M., Vijay, K. M., & Douglas, B. W. (1999). Spectrum estimation and modeling. Digital Signal Processing Handbook, CRC Press LLC.

Gupta, H. R., Mehra, R., & Batan, S. (2013). Power spectrum estimation using Welch method for various window techniques. International Journal of Scientific Research Engineering & Technology, 2(6), 389-392.

Hasselmann, K.; Barnett, T. P.; Bouws, E.; Carlson, H.; Cartwright, D. E.; Enke, K.; Ewing, J. A.; Gienapp, H.; Hasselmann, D. E.; Kruseman, P.; Meerbrug, A.; Muller, P.; Olbers, D. J.; Richter, K.; Sell, W., and Walden, H., 1973. Measurements of wind-wave growth and swell decay during the Joint North Sea Wave Project (JONSWAP). Deutsche Hydrographische Zeitschrift A80(12), 95p.

Karuna, M., & Mahalakshmi, B. (2015), Comparability of Power Spectral Density Estimation of EMG Signals Using Non-Parametric Methods, International Journal of Science and Research, Vol. 4, Issue 9, pp.189-192.

Rahi P K, Mehra R (2014), Analysis of power spectrum estimation using Welch method for various window techniques, International Journal of Emerging Technologies and Engineering, Vol. 2, Issue 6, pp.106-109.

Shradhanjali, A., Chowdhury, S., & Kumar, N. (2013). Power spectral density estimation of EMG signals using parametric and non-parametric approach. Global Advanced Research Journal of Engineering, Technology and Innovation, 2(4), 111-7.

Smith, S. W. (1997). The scientist and engineer's guide to digital signal processing.

Welch, P. (1967). The use of fast Fourier transform for the estimation of power spectra: a method based on time averaging over short, modified periodograms. IEEE Transactions on audio and electroacoustics, 15(2), 70-73.

Chapter 6

Ardhuin, F., Rogers, E., Babanin, A. V., Filipot, J. F., Magne, R., Roland, A., ... & Collard, F. (2010). Semiempirical dissipation source functions for ocean waves. Part I: Definition, calibration, and validation. Journal of Physical Oceanography, 40(9), 1917-1941.

Bergdahl, L. (2009). Comparison of measured shallow-water wave spectra with theoretical spectra. In Proc. 8th European Wave and Tidal Energy conference (pp. 100-105).

Bishop, C. T., & Donelan, M. A. (1987). Measuring waves with pressure transducers. Coastal Engineering, 11(4), 309-328.

Bouws, E.; Günther, H.; Rosenthal, W., and Vincent, C.L., (1985). Similarity of the wind wave spectrum in finite depth water: 1. Spectral form. Journal of Geophysical Research: Oceans, 90(C1), 975-986.

Cavaleri, L., (1980). Wave measurement using pressure transducer. Oceanologica Acta, 3(3), 339-346.

Hasselmann, K.; Barnett, T. P.; Bouws, E.; Carlson, H.; Cartwright, D. E.; Enke, K.; Ewing, J. A.; Gienapp, H.; Hasselmann, D. E.; Kruseman, P.; Meerbrug, A.; Muller, P.; Olbers, D. J.; Richter, K.; Sell, W., and Walden, H., 1973. Measurements of wind-wave growth and swell decay during the Joint North Sea Wave Project (JONSWAP). Deutsche Hydrographische Zeitschrift A80(12), 95p.

Holthuijsen, L. H. (2007). Waves in oceanic and coastal waters. Cambridge University Press.

Huang, M. C., & Tsai, C. H. (2008). Pressure transfer function in time and time-frequency domains. Ocean Engineering, 35(11), 1203-1210.

Hughes, S. A. (1984). The TMA shallow-water spectrum description and applications (No. CERC-TR-84-7). COASTAL ENGINEERING RESEARCH CENTER VICKSBURG MS.

Jones, N. L., & Monismith, S. G. (2007). Measuring short-period wind waves in a tidally forced environment with a subsurface pressure gauge. Limnology and Oceanography: Methods, 5(10), 317-327.

Kaihatu, J. M., Veeramony, J., Edwards, K. L. & Kirby, J. T. (2007) Asymptotic behaviour of frequency and wave number spectra of nearshore shoaling and breaking waves. J. Geophys. Res. 112, C06016

Kitaigordskii, S. A., Krasitskii, V. P., & Zaslavskii, M. M. (1975). On Phillips' theory of equilibrium range in the spectra of wind-generated gravity waves. Journal of Physical Oceanography, 5(3), 410-420.

Kitaigorodskii, S. A. (1983). On the theory of the equilibrium range in the spectrum of wind-generated gravity waves. Journal of Physical Oceanography, 13(5), 816-827.

Kuo, Y. Y., & Chiu, Y. F. (1994). Transfer function between wave height and wave pressure for progressive waves. Coastal Engineering, 23(1), 81-93.

Miller, H. C., & Vincent, C. L. (1990). FRF spectrum: TMA with Kitaigorodskii's f-4 scaling. Journal of Waterway, Port, Coastal, and Ocean Engineering, 116(1), 57-78.

Phillips, O. M. (1958). The equilibrium range in the spectrum of wind-generated waves. Journal of Fluid Mechanics, 4(04), 426-434.

Pierson, W. J., & Moskowitz, L. (1964). A proposed spectral form for fully developed wind seas based on the similarity theory of SA Kitaigorodskii. Journal of geophysical research, 69(24), 5181-5190.

Siadatmousavi, S. M., Jose, F., & Stone, G. W. (2012). On the importance of high frequency tail in third generation wave models. Coastal Engineering.

Smith, M. J., Stevens, C. L., Gorman, R. M., McGregor, J. A., & Neilson, C. G. (2001). Wind-wave development across a large shallow intertidal estuary: A case study of Manukau Harbour, New Zealand. New Zealand Journal of Marine and Freshwater Research, 35(5), 985-1000.

Smith, J. M. (2002). Wave pressure gauge analysis with current. Journal of waterway, port, coastal, and ocean engineering, 128(6), 271-275.

Thompson, E. F., and Vincent, C. L. (1983). Prediction of wave height in shallow water. Proceedings of Coastal Structures 1983, American Society of Civil Engineers, 1000–1008.

Thornton, E. B. (1977). Rederivation of the saturation range in the frequency spectrum of wind-generated gravity waves. Journal of Physical Oceanography, 7(1), 137-140.

Townsend, M., & Fenton, J. D. (1996). A comparison of analysis methods for wave pressure data. Coastal Engineering Proceedings, 1(25).

Trowbridge, J., & Elgar, S. (2001). Turbulence Measurements in the Surf Zone*. Journal of Physical Oceanography, 31(8), 2403-2417.

Tsai, C. H., Huang, M. C., Young, F. J., Lin, Y. C., & Li, H. W. (2005). On the recovery of surface wave by pressure transfer function. Ocean Engineering, 32(10), 1247-1259.

Tucker, M. J. (1994). Nearshore waveheight during storms. Coastal Engineering, 24(1-2), 111-136.

Van Rijn, L. C., Grasmeijer, B. T., & Ruessink, B. G. (2000). Measurement errors of instruments for velocity, wave heigt, sand concentration and bed levels in field conditions. Deltares (WL)-Utrecht University.

Chapter 7

Ardhuin, F., Rogers, E., Babanin, A. V., Filipot, J. F., Magne, R., Roland, A., ... & Collard, F. (2010). Semiempirical dissipation source functions for ocean waves. Part I: Definition, calibration, and validation. Journal of Physical Oceanography, 40(9), 1917-1941.

Bergdahl, L. (2009). Comparison of measured shallow-water wave spectra with theoretical spectra. In Proc. 8th European Wave and Tidal Energy conference (pp. 100-105).

Bouws, E.; Günther, H.; Rosenthal, W., and Vincent, C.L., (1985). Similarity of the wind wave spectrum in finite depth water: 1. Spectral form. Journal of Geophysical Research: Oceans, 90(C1), 975-986.

Ciavola, P., & Coco, G. (2017). Coastal Storms: Processes and Impacts. John Wiley & Sons.

Dronkers, J. (2005). Dynamics of coastal systems (Vol. 25). World Scientific.

Hasselmann, K.; Barnett, T. P.; Bouws, E.; Carlson, H.; Cartwright, D. E.; Enke, K.; Ewing, J. A.; Gienapp, H.; Hasselmann, D. E.; Kruseman, P.; Meerbrug, A.; Muller, P.; Olbers, D. J.; Richter, K.; Sell, W., and Walden, H., 1973. Measurements of wind-wave growth and swell decay during the Joint North Sea Wave Project (JONSWAP). Deutsche Hydrographische Zeitschrift A80(12), 95p.

Holthuijsen, L. H. (2007). Waves in oceanic and coastal waters. Cambridge University Press.

Hughes, S. A. (1984). The TMA shallow-water spectrum description and applications (No. CERC-TR-84-7). COASTAL ENGINEERING RESEARCH CENTER VICKSBURG MS.

Jones, N. L., & Monismith, S. G. (2007). Measuring short-period wind waves in a tidally forced environment with a subsurface pressure gauge. Limnology and Oceanography: Methods, 5(10), 317-327.

Kaihatu, J. M., Veeramony, J., Edwards, K. L. & Kirby, J. T. (2007) Asymptotic behaviour of frequency and wave number spectra of nearshore shoaling and breaking waves. J. Geophys. Res. 112, C06016

Karimpour, A., Chen, Q., & Twilley, R. R. (2016). A field study of how wind waves and currents may contribute to the deterioration of saltmarsh fringe. Estuaries and Coasts, 39(4), 935-950.

Kitaigordskii, S. A., Krasitskii, V. P., & Zaslavskii, M. M. (1975). On Phillips' theory of equilibrium range in the spectra of wind-generated gravity waves. Journal of Physical Oceanography, 5(3), 410-420.

Kitaigorodskii, S. A. (1983). On the theory of the equilibrium range in the spectrum of wind-generated gravity waves. Journal of Physical Oceanography, 13(5), 816-827.

Miller, H. C., & Vincent, C. L. (1990). FRF spectrum: TMA with Kitaigorodskii's f-4 scaling. Journal of Waterway, Port, Coastal, and Ocean Engineering, 116(1), 57-78.

Nielsen, P. (1992). Coastal bottom boundary layers and sediment transport (Vol. 4). World Scientific Publishing Co Inc.

Phillips, O. M. (1958). The equilibrium range in the spectrum of wind-generated waves. Journal of Fluid Mechanics, 4(04), 426-434.

Pierson, W. J., & Moskowitz, L. (1964). A proposed spectral form for fully developed wind seas based on the similarity theory of SA Kitaigorodskii. Journal of geophysical research, 69(24), 5181-5190.

Reeve, D., Chadwick, A., & Fleming, C. (2012). Coastal engineering: processes, theory and design practice. CRC Press.

Siadatmousavi, S. M., Jose, F., & Stone, G. W. (2012). On the importance of high frequency tail in third generation wave models. Coastal Engineering.

Smith, J. M. (2002). Wave pressure gauge analysis with current. Journal of waterway, port, coastal, and ocean engineering, 128(6), 271-275.

Thompson, E. F., and Vincent, C. L. (1983). Prediction of wave height in shallow water. Proceedings of Coastal Structures 1983, American Society of Civil Engineers, 1000–1008.

Thornton, E. B. (1977). Rederivation of the saturation range in the frequency spectrum of wind-generated gravity waves. Journal of Physical Oceanography, 7(1), 137-140.

Tucker, M. J. (1994). Nearshore waveheight during storms. Coastal Engineering, 24(1-2), 111-136.

Wiberg, P. L., & Sherwood, C. R. (2008). Calculating wave-generated bottom orbital velocities from surface-wave parameters. Computers & Geosciences, 34(10), 1243-1262.

Chapter 8

Banner, M. L. (1990). Equilibrium spectra of wind waves. Journal of Physical Oceanography, 20(7), 966-984.

Beji, S. (2013). Improved explicit approximation of linear dispersion relationship for gravity waves. Coastal Engineering, 73, 11-12.

Benoit M., Frigaard P. and Schaffer H., (1997), Analysing multidirectional wave spectra: A tentative classification of available methods, Proc. IAHR Seminar on Multidirectional Waves and Their Interaction with Structures, 27th IAHR Congress, ed. E. Mansard, San Francisco, USA (1997).

Bretschneider, C. L. (1959). Wave variability and wave spectra for wind-generated gravity waves (No. TM-118). CORPS OF ENGINEERS WASHINGTON DC BEACH EROSION BOARD.

Bouws, E.; Günther, H.; Rosenthal, W., and Vincent, C.L., (1985). Similarity of the wind wave spectrum in finite depth water: 1. Spectral form. Journal of Geophysical Research: Oceans, 90(C1), 975-986.

Casas-Prat, M., & Holthuijsen, L. H. (2010). Short-term statistics of waves observed in deep water. Journal of Geophysical Research: Oceans, 115(C9).

Chakrabarti, S. K. (1987). Hydrodynamics of offshore structures. WIT press.

Dean, R. G., & Dalrymple, R. A. (1991). Water wave mechanics for engineers and scientists (Vol. 2). World Scientific Publishing Company.

Deo, M. C., Gondane, D. S., & Sanil Kumar, V. (2002). Analysis of wave directional spreading using neural networks. Journal of waterway, port, coastal, and ocean engineering, 128(1), 30-37.

Donelan, M. A., Hamilton, J., & Hui, W. (1985). Directional spectra of wind-generated waves. Philosophical Transactions of the Royal Society of London A: Mathematical, Physical and Engineering Sciences, 315(1534), 509-562.

Donelan, M., Babanin, A., Sanina, E., & Chalikov, D. (2015). A comparison of methods for estimating directional spectra of surface waves. Journal of Geophysical Research: Oceans, 120(7), 5040-5053.

Earle, M. D., McGehee, D., & Tubman, M. (1995). Field Wave Gaging Program, Wave Data Analysis Standard (No. WES/IR/CERC-95-2). ARMY ENGINEER WATERWAYS EXPERIMENT STATION VICKSBURG MS.

Earle, M. D., Steele, K. E., & Wang, D. W. C. (1999). Use of advanced directional wave spectra analysis methods. Ocean engineering, 26(12), 1421-1434.

Ewans, K. C. (1998). Observations of the directional spectrum of fetch-limited waves. Journal of Physical Oceanography, 28(3), 495-512.

Fenton, J. D., & McKee, W. D. (1990). On calculating the lengths of water waves. Coastal Engineering, 14(6), 499-513.

Gilhousen, D.B. and R. Hervey, (2001). Improved Estimates of Swell from Moored Buoys. Proceedings of the Fourth International Symposium WAVES 2001, ASCE: Alexandria, VA, pp. 387-393.

Grosskopf, W., Aubrey, D., Mattie, M., & Mathiesen, M. (1983). Field intercomparison of nearshore directional wave sensors. IEEE Journal of Oceanic Engineering, 8(4), 254-271.

Goda, Y. (1999). A comparative review on the functional forms of directional wave spectrum. Coastal Engineering Journal, 41(01), 1-20.

Goda, Y. (2010). Random seas and design of maritime structures (Vol. 33). World Scientific Publishing Company.

Guo, J. (2002). Simple and explicit solution of wave dispersion equation. Coastal Engineering, 45(2), 71-74.

Hasselmann, K.; Barnett, T. P.; Bouws, E.; Carlson, H.; Cartwright, D. E.; Enke, K.; Ewing, J. A.; Gienapp, H.; Hasselmann, D. E.; Kruseman, P.; Meerbrug, A.; Muller, P.; Olbers, D. J.; Richter, K.; Sell, W., and Walden, H., (1973). Measurements of wind-wave growth and swell decay during the Joint North Sea Wave Project (JONSWAP). Deutsche Hydrographische Zeitschrift A80(12), 95p.

Hasselmann, D. E., Dunckel, M., & Ewing, J. A. (1980). Directional wave spectra observed during JONSWAP 1973. Journal of physical oceanography, 10(8), 1264-1280.

Herbers, T. H. C., Elgar, S., & Guza, R. T. (1999). Directional spreading of waves in the nearshore. Journal of Geophysical Research: Oceans, 104(C4), 7683-7693.

Holthuijsen, L. H. (2007). Waves in oceanic and coastal waters. Cambridge University Press.

Hunt, J. N. (1979). Direct solution of wave dispersion equation. Journal of the Waterway Port Coastal and Ocean Division, 105(4), 457-459.

Hwang, P. A., & Wang, D. W. (2001). Directional distributions and mean square slopes in the equilibrium and saturation ranges of the wave spectrum. Journal of physical oceanography, 31(5), 1346-1360.

Hwang, P. A., Ocampo-Torres, F. J., & García-Nava, H. (2012). Wind sea and swell separation of 1D wave spectrum by a spectrum integration method. Journal of Atmospheric and Oceanic Technology, 29(1), 116-128.

Jones, N. L., & Monismith, S. G. (2007). Measuring short-period wind waves in a tidally forced environment with a subsurface pressure gauge. Limnology and Oceanography: Methods, 5(10), 317-327.

Kim, Y. C. (2010). Handbook of coastal and ocean engineering. World Scientific.

Kumar, V. S., & Anand, N. M. (2004). Variations in wave direction estimated using first and second order Fourier coefficients. Ocean engineering, 31(17), 2105-2119.

Liu, Y., Yan, X. H., Liu, W. T., & Hwang, P. A. (1997). The probability density function of ocean surface slopes and its effects on radar backscatter. Journal of physical oceanography, 27(5), 782-797.

Lounget-Higgins, M. S. (1952). On the Statistical Distribution of the Heights of Sea Wave. J. of Marine Research, 11(3), 245-266.

Longuet-Higgins, M. S., Cartwright, D. E., & Smith, N. D. (1963). Observations of the directional spectrum of sea waves using the motions of a floating buoy.

Longuet-Higgins, M. S. (1975). On the joint distribution of the periods and amplitudes of sea waves. Journal of Geophysical Research, 80(18), 2688-2694.

Longuet-Higgins, M. S. (1983). On the joint distribution of wave periods and amplitudes in a random wave field. In Proceedings of the Royal Society of London A: Mathematical, Physical and Engineering Sciences (Vol. 389, No. 1797, pp. 241-258). The Royal Society.

Massel, S. R., & Brinkman, R. M. (1998). On the determination of directional wave spectra for practical applications. Applied ocean research, 20(6), 357-374.

McCormick, M. E. (2009). Ocean engineering mechanics: with applications. Cambridge University Press.

Memos, C., & Ziros, A. (2003). Field measurements of wave directionality in water of finite depth. Ocean and Polar Research, 25(4), 437-446.

Mitsuyasu, H., Tasai, F., Suhara, T., Mizuno, S., Ohkusu, M., Honda, T., & Rikiishi, K. (1975). Observations of the directional spectrum of ocean WavesUsing a cloverleaf buoy. Journal of Physical Oceanography, 5(4), 750-760.

Pedersen, T., Lohrmann, A., & Krogstad, H. E. (2005). Wave measurements from a subsurface platform. Proceedings WAVES 2005.

Pierson, W. J., Neumann, G., & James, R. W. (1955). Practical methods for observing and forecasting ocean waves by means of wave spectra and statistics. Publication 603, U.S. Navy Hydrographic Office, 284 pp.

Portilla, J., Ocampo-Torres, F. J., & Monbaliu, J. (2009). Spectral partitioning and identification of wind sea and swell. Journal of atmospheric and oceanic technology, 26(1), 107-122.

Reeve, D., Chadwick, A., & Fleming, C. (2012). Coastal engineering: processes, theory and design practice. CRC Press.

Smith, J. M. (2002). Wave pressure gauge analysis with current. Journal of waterway, port, coastal, and ocean engineering, 128(6), 271-275.

Soares, C. G., & Carvalho, A. N. (2003). Probability distributions of wave heights and periods in measured combined sea-states from the Portuguese coast. TRANSACTIONS-AMERICAN SOCIETY OF MECHANICAL ENGINEERS JOURNAL OF OFFSHORE MECHANICS AND ARCTIC ENGINEERING, 125(3), 198-204.

Sorensen, R. M. (2005). Basic coastal engineering (Vol. 10). Springer Science & Business Media.

Soulsby, R. L. (2006). Simplified calculation of wave orbital velocities.

St Denis, M., & Pierson Jr, W. J. (1953). On the motions of ships in confused seas. NEW YORK UNIV BRONX SCHOOL OF ENGINEERING AND SCIENCE.

Vatankhah, A. R., & Aghashariatmadari, Z. (2013). Improved explicit approximation of linear dispersion relationship for gravity waves: A discussion. Coastal engineering, 78, 21-22.

Wang, D., & Gilhousen, D. (1998). Separation of seas and swells from NDBC buoy wave data. In Fifth Int. Workshop on Wave Hindcasting and Forecasting (pp. 155-162). Environment Canada Melbourne, Florida, FL, ASCE.

Work, P. A. (2008). Nearshore directional wave measurements by surface-following buoy and acoustic Doppler current profiler. Ocean Engineering, 35(8), 727-737.

Xu, D., Li, X., Zhang, L., Xu, N., & Lu, H. (2004). On the distributions of wave periods, wavelengths, and amplitudes in a random wave field. Journal of Geophysical Research: Oceans, 109(C5).

Yamaguchi, M., & Nonaka, H. (2007). Comparative study of explicit solutions to wave dispersion equation. Annual Journal of Eng., Ehime Univ., VI, Mar., 2007.

Yamaguchi, M., Nonaka, H., & Hatada, Y. (2014). Comparative Study of Explicit Solutions to Wave Dispersion Relationship (2). Annual Journal of Eng., Ehime Univ., XIII, Mar., 2014.

You, Z. J. (2008). A close approximation of wave dispersion relation for direct calculation of wavelength in any coastal water depth. Applied Ocean Research, 30(2), 113-119.

Young, I. R. (1999). Wind generated ocean waves (Vol. 2). Elsevier.

Yu, J. (2013). Approximations of the Dispersion Relationship of Water Waves. Journal of Engineering Mechanics, 140(1), 233-235.

Zhang, H. D., Cherneva, Z., & Soares, C. G. (2013). Joint distributions of wave height and period in laboratory generated nonlinear sea states. Ocean Engineering, 74, 72-80.

Chapter 9

Amante, C., & Eakins, B. W. (2009). ETOPO1 1 Arc-Minute Global Relief Model: Procedures, Data Sources and Analysis. NOAA Technical Memorandum NESDIS NGDC-24. National Geophysical Data Center, NOAA. doi:10.7289/V5C8276M

American Society of Civil Engineers, ASCE-7. (2005). Minimum design loads for buildings and other structures (Vol. 7). Amer Society of Civil Engineers.

Ashcroft, J. (1994). The relationship between the gust ratio, terrain roughness, gust duration and the hourly mean wind speed. Journal of Wind Engineering and Industrial Aerodynamics, 53(3), 331-355.

Atkinson, G. D., & Holliday, C. R. (1977). Tropical cyclone minimum sea level pressure/maximum sustained wind relationship for the western north Pacific. Monthly Weather Review, 105(4), 421-427.

Bailey, D. T. (2000). Meteorological monitoring guidance for regulatory modeling applications. DIANE Publishing.

Bardal, L. M., & Sætran, L. R. (2016). Wind gust factors in a coastal wind climate. Energy Procedia, 94, 417-424.

Bennett, N. D., Croke, B. F., Guariso, G., Guillaume, J. H., Hamilton, S. H., Jakeman, A. J., Marsili-Libelli, S., Newham, L. T., Norton, J. P., Perrin, C. and Pierce, S. A. (2013). Characterising performance of environmental models. Environmental Modelling & Software, 40, 1-20.

Bergdahl, L. (2009). Comparison of measured shallow-water wave spectra with theoretical spectra. In Proc. 8th European Wave and Tidal Energy conference (pp. 100-105).

Bishop, C. T. (1983). Comparison of manual wave prediction models. Journal of Waterway, Port, Coastal, and Ocean Engineering, 109(1), 1-17.

Booij, N. R. R. C., Ris, R. C., & Holthuijsen, L. H. (1999). A third-generation wave model for coastal regions: 1. Model description and validation. Journal of geophysical research: Oceans, 104(C4), 7649-7666.

Bouws, E.; Günther, H.; Rosenthal, W., and Vincent, C.L., (1985). Similarity of the wind wave spectrum in finite depth water: 1. Spectral form. Journal of Geophysical Research: Oceans, 90(C1), 975-986.

Bretschneider, C. L. (1952). Revised wave forecasting relationships. Coastal Engineering Proceedings, 1(2), 1.

Bretschneider, C. L. (1958). Revisions in wave forecasting: deep and shallow water. Coastal Engineering Proceedings, 1(6), 3.

Bretschneider, C. L. (1959). Wave variability and wave spectra for wind-generated gravity waves (No. TM-118). CORPS OF ENGINEERS WASHINGTON DC BEACH EROSION BOARD.

Bretschneider, C. L. (1965). Generation of waves by wind. State of the art (No. NESCO-SN-134-6). NATIONAL ENGINEERING SCIENCE CO WASHINGTON DC.

Bretschneider, C. L. (1972, January). A non-dimensional stationary hurricane wave model. In Offshore Technology Conference. Offshore Technology Conference.

Cao, S., Tamura, Y., Kikuchi, N., Saito, M., Nakayama, I., & Matsuzaki, Y. (2009). Wind characteristics of a strong typhoon. Journal of wind engineering and industrial aerodynamics, 97(1), 11-21.

Carter, D. J. T. (1982). Prediction of wave height and period for a constant wind velocity using the JONSWAP results. Ocean Engineering, 9(1), 17-33.

Charnock, H. (1955). Wind stress on a water surface. Quarterly Journal of the Royal Meteorological Society, 81(350), 639-640.

Chavas, D. R., & Emanuel, K. A. (2010). A QuikSCAT climatology of tropical cyclone size. Geophysical Research Letters, 37(18).

Chen, C., Beardsley, R.C., Cowles, G., Qi, J., Lai, Z., Gao, G., Stuebe, D., Xu, Q., Xue, P., Ge, J. and Ji, R., (2012). An Unstructured-grid, Finite-volume Community Ocean Model: FVCOM User Manual. Sea Grant College Program, Massachusetts Institute of Technology.

Cheung, K. F., Tang, L., Donnelly, J. P., Scileppi, E. M., Liu, K. B., Mao, X. Z., ... & Murnane, R. J. (2007). Numerical modeling and field evidence of coastal overwash in southern New England from Hurricane Bob and implications for paleotempestology. Journal of Geophysical Research: Earth Surface, 112(F3).

CIRIA, Construction Industry Research, Information Association, Civieltechnisch Centrum Uitvoering Research en Regelgeving (Netherlands), & Centre d'études maritimes et fluviales (France). (2007). The Rock Manual: The use of rock in hydraulic engineering (Vol. 683). Ciria.

Cook, N. J. (1985). The designer's guide to wind loading on building structures. Part I: Background, damage survey, wind data, and structural classification. Building Research Establishment, Watford.

Davenport, A.G. (1960). A Rationale for the Determination of Design Wind Velocities. Proceedings ASCE Structural Division, 86: 39-66.

Deacon, E. L. (1965). Wind gust speed: averaging time relationship. Aust. Meteorol. Mag, 51, 11-14.

DeMaria, M. (1987). Tropical cyclone track prediction with a barotropic spectral model. Monthly weather review, 115(10), 2346-2357.

Department of the Army, Waterways Experiment Station, Corps of Engineers, and Coastal Engineering Research Center (1984), Shore Protection Manual, Washington, D.C., vol. 1, 4th ed., 532 pp.

Donelan, M. A., Hamilton, J., & Hui, W. (1985). Directional spectra of wind-generated waves. Philosophical Transactions of the Royal Society of London. Series A, Mathematical and Physical Sciences, 509-562.

Durst, C. S. (1960). Wind speeds over short periods of time. Meteor. Mag, 89(1056), 181-187.

Emanuel, K., & Rotunno, R. (2011). Self-stratification of tropical cyclone outflow. Part I: Implications for storm structure. Journal of the Atmospheric Sciences, 68(10), 2236-2249.

Emeis, S. (2012). Wind energy meteorology: atmospheric physics for wind power generation. Springer Science & Business Media.

Engineering Sciences Data Unit, ESDU, (2002), Strong winds in the atmospheric boundary layer. Part 2: discrete gust speeds. Engineering Science Data Unit, Item No. 83045, London.

Etemad-Shahidi, A., Kazeminezhad, M. H., & Mousavi, S. J. (2009). On the prediction of wave parameters using simplified methods. Journal of Coastal Research, 505-509.

Fairall, C. W., Bradley, E. F., Rogers, D. P., Edson, J. B., & Young, G. S. (1996). Bulk parameterization of air-sea fluxes for tropical ocean-global atmosphere coupled-ocean atmosphere response experiment. Journal of Geophysical Research: Oceans, 101(C2), 3747-3764.

Farrugia, P. S., & Micallef, A. (2006). Comparative analysis of estimators for wind direction standard deviation. Meteorological Applications, 13(1), 29-41.

Fisher, A. W., Sanford, L. P., & Suttles, S. E. (2015). Wind stress dynamics in Chesapeake Bay: Spatiotemporal variability and wave dependence in a fetch-limited environment. Journal of Physical Oceanography, 45(10), 2679-2696.

Garratt, J. R. (1977). Review of drag coefficients over oceans and continents. Monthly weather review, 105(7), 915-929.

Goda, Y. (2003). Revisiting Wilson's formulas for simplified wind-wave prediction. Journal of waterway, port, coastal, and ocean engineering, 129(2), 93-95.

Graham and Numm (1959), Meteorological Conditions Pertinent to Standard Project Hurricane, Atlantic and Gulf Coasts of United States. National Hurricane Research Project. U.S. Weather Service, Report no. 33.

Greenway, M. E. (1979). An analytical approach to wind velocity gust factors. Journal of Wind Engineering and Industrial Aerodynamics, 5(1-2), 61-91.

Handbook, A. S. H. R. A. E. (2009). Fundamentals, Airflow around buildings. ASHRAE Handbook, Fundamentals, (2009).

Harmel, R. D., Smith, P. K., & Migliaccio, K. W. (2010). Modifying goodness-of-fit indicators to incorporate both measurement and model uncertainty in model calibration and validation. Transactions of the ASAE (American Society of Agricultural Engineers), 53(1), 55.

Harper, B. A., & Holland, G. J. (1999, January). An updated parametric model of the tropical cyclone. In Proc. 23rd Conf. Hurricanes and Tropical Meteorology.

Harper, B. A., Kepert, J. D., & Ginger, J. D. (2010). Guidelines for converting between various wind averaging periods in tropical cyclone conditions. World Meteorological Organization, WMO/TD, 1555.

Harper, B. A. (2013, July). Best practice in tropical cyclone wind hazard modelling: In search of data and emptying the skeleton cupboard. In Proceedings of the 16th Australasian Wind Engineering Society Workshop, Brisbane, Qld, Australia (pp. 18-19).

Hasselmann, K.; Barnett, T. P.; Bouws, E.; Carlson, H.; Cartwright, D. E.; Enke, K.; Ewing, J. A.; Gienapp, H.; Hasselmann, D. E.; Kruseman, P.; Meerbrug, A.; Muller, P.; Olbers, D. J.; Richter, K.; Sell, W., and Walden, H., (1973). Measurements of wind-wave growth and swell decay during the Joint North Sea Wave Project (JONSWAP). Deutsche Hydrographische Zeitschrift A80(12), 95p.

He, Y. C., Chan, P. W., & Li, Q. S. (2014). Standardization of raw wind speed data under complex terrain conditions: a data-driven scheme. Journal of Wind Engineering and Industrial Aerodynamics, 131, 12-30.

Holland, G. J. (1980). An analytic model of the wind and pressure profiles in hurricanes. Monthly weather review, 108(8), 1212-1218.

Holland, G. (2008). A revised hurricane pressure–wind model. Monthly Weather Review, 136(9), 3432-3445.

Holland, G. J., Belanger, J. I., & Fritz, A. (2010). A revised model for radial profiles of hurricane winds. Monthly Weather Review, 138(12), 4393-4401.

Holmes, J. D., Allsop, A. C., & Ginger, J. D. (2014). Gust durations, gust factors and gust response factors in wind codes and standards. Wind & Structures, 19, 339-352.

Holmes, J. D. (2015). Wind loading of structures. CRC press.

Holthuijsen, L. H. (2007). Waves in oceanic and coastal waters. Cambridge University Press.

Hu, K., Chen, Q., & Kimball, S. K. (2012). Consistency in hurricane surface wind forecasting: an improved parametric model. Natural hazards, 61(3), 1029-1050.

Hughes, S. A. (1984). The TMA shallow-water spectrum description and applications (No. CERC-TR-84-7). COASTAL ENGINEERING RESEARCH CENTER VICKSBURG MS.

Hwang, P. A., & Wang, D. W. (2004). Field measurements of duration-limited growth of wind-generated ocean surface waves at young stage of development. Journal of Physical Oceanography, 34(10), 2316-2326.

Ishizaki, H. (1983). Wind profiles, turbulence intensities and gust factors for design in typhoon-prone regions. Journal of Wind Engineering and Industrial Aerodynamics, 13(1-3), 55-66.

ISSC, (1964). Proceedings of the 2nd International Ship Structures Congress, Delft, The Netherlands.

Jelesnianski, C. P., Chen, J., & Shaffer, W. A. (1992). SLOSH: Sea, lake, and overland surges from hurricanes (Vol. 48). US Department of Commerce, National Oceanic and Atmospheric Administration, National Weather Service.

Kahma, K. K., & Calkoen, C. J. (1992). Reconciling discrepancies in the observed growth of wind-generated waves. Journal of Physical Oceanography, 22(12), 1389-1405.

Karimpour, A., Kaye, N. B., & Baratian-Ghorghi, Z. (2012). Modeling the neutrally stable atmospheric boundary layer for laboratory scale studies of the built environment. Building and Environment, 49, 203-211.

Karimpour, A., & Chen, Q. (2016). A Simplified Parametric Model for Fetch-Limited Peak Wave Frequency in Shallow Estuaries. Journal of Coastal Research, 32(4), 954.

Karimpour, A., Chen, Q., & Twilley, R. R. (2017). Wind Wave Behavior in Fetch and Depth Limited Estuaries. Scientific reports, 7, 40654.

Kirby, J. T., Wei, G., Chen, Q., Kennedy, A. B., & Dalrymple, R. A. (1998). FUNWAVE 1.0: fully nonlinear Boussinesq wave model-Documentation and user's manual. research report NO. CACR-98-06.

Kitaigordskii, S. A., Krasitskii, V. P., & Zaslavskii, M. M. (1975). On Phillips' theory of equilibrium range in the spectra of wind-generated gravity waves. Journal of Physical Oceanography, 5(3), 410-420.

Krayer, W. R., & Marshall, R. D. (1992). Gust factors applied to hurricane winds. Bulletin of the American Meteorological Society, 73(5), 613-618.

Kwon, D. K., & Kareem, A. (2013). Comparative study of major international wind codes and standards for wind effects on tall buildings. Engineering Structures, 51, 23-35.

Letchford, C. W., & Zachry, B. C. (2009). On wind, waves, and surface drag. In 5th European and African conference on wind engineering, Florence, Italy.

Liu, Q., Babanin, A., Fan, Y., Zieger, S., Guan, C., & Moon, I. J. (2017). Numerical simulations of ocean surface waves under hurricane conditions: Assessment of existing model performance. Ocean Modelling, 118, 73-93.

Liu, Y., Li, S., Yi, Q., & Chen, D. (2017). Wind Profiles and Wave Spectra for Potential Wind Farms in South China Sea. Part II: Wave Spectrum Model. Energies, 10(1), 127.

Mahadevan, A. (2006). Modeling vertical motion at ocean fronts: Are nonhydrostatic effects relevant at submesoscales?. Ocean Modelling, 14(3), 222-240.

Martino, C. D., Cheung, K. F., Phadke, A. C., & Houston, S. H. (2001, January). Modeling of hurricane waves in Hawaiian waters. In The Eleventh International Offshore and Polar Engineering Conference. International Society of Offshore and Polar Engineers.

Masters, F. J., Vickery, P. J., Bacon, P., & Rappaport, E. N. (2010). Toward objective, standardized intensity estimates from surface wind speed observations. Bulletin of the American Meteorological Society, 91(12), 1665-1681.

Mathews, S. T. (1972). A Critical Review of the 12th ITTC Wave Spectrum Recommendations. Report of Seakeeping Committee, Appendix, 9, 973-986.

McCormick, M. E. (2009). Ocean engineering mechanics: with applications. Cambridge University Press.

Mitsuta, Y., & Tsukamoto, O. (1989). Studies on spatial structure of wind gust. Journal of Applied Meteorology, 28(11), 1155-1160.

Mitsuyasu, H. (1973). The one-dimensional wave spectra at limited fetch. In Coastal Engineering 1972 (pp. 289-306).

Mitsuyasu, H., Tasai, F., Suhara, T., Mizuno, S., Ohkusu, M., Honda, T., & Rikiishi, K. (1980). Observation of the power spectrum of ocean waves using a cloverleaf buoy. Journal of Physical Oceanography, 10(2), 286-296.

Mori, Y. (1986). Evaluation of several "single-pass" estimators of the mean and the standard deviation of wind direction. Journal of climate and applied meteorology, 25(10), 1387-1397.

Moriasi, D. N., Arnold, J. G., Van Liew, M. W., Bingner, R. L., Harmel, R. D., & Veith, T. L. (2007). Model evaluation guidelines for systematic quantification of accuracy in watershed simulations. Trans. Asabe, 50(3), 885-900.

Ochi, M. K., & Hubble, E. N. (1977). Six-parameter wave spectra. In Coastal Engineering 1976 (pp. 301-328).

Ochi, M. K. (1982). Stochastic analysis and probabilistic prediction of random seas. Advances in Hydroscience, 13, 217-375.

Ochi, M. K. (2005). Ocean waves: the stochastic approach (Vol. 6). Cambridge University Press.

Orton, P. M., Hall, T. M., Talke, S. A., Blumberg, A. F., Georgas, N., & Vinogradov, S. (2016). A validated tropical-extratropical flood hazard assessment for New York Harbor. Journal of Geophysical Research: Oceans, 121(12), 8904-8929.

Phadke, A. C., Martino, C. D., Cheung, K. F., & Houston, S. H. (2003). Modeling of tropical cyclone winds and waves for emergency management. Ocean Engineering, 30(4), 553-578.

Phillips, O. M. (1958). The equilibrium range in the spectrum of wind-generated waves. Journal of Fluid Mechanics, 4(4), 426-434.

Pierson, W. J., & Moskowitz, L. (1964). A proposed spectral form for fully developed wind seas based on the similarity theory of SA Kitaigorodskii. Journal of geophysical research, 69(24), 5181-5190.

Powell, M. D., & Black, P. G. (1990). The relationship of hurricane reconnaissance flight-level wind measurements to winds measured by NOAA's oceanic platforms. Journal of Wind Engineering and Industrial Aerodynamics, 36, 381-392.

Powell, M. D., & Houston, S. H. (1996). Hurricane Andrew's landfall in South Florida. Part II: Surface wind fields and potential real-time applications. Weather and Forecasting, 11(3), 329-349.

Reeve, D., Chadwick, A., & Fleming, C. (2012). Coastal engineering: processes, theory and design practice. CRC Press.

Saville Jr, T. (1954). The effect of fetch width on wave generation (No. TM70). COASTAL ENGINEERING RESEARCH CENTER VICKSBURG MS.

Shchepetkin, A. F., & McWilliams, J. C. (2005). The regional oceanic modeling system (ROMS): a split-explicit, free-surface, topography-following-coordinate oceanic model. Ocean Modelling, 9(4), 347-404.

Shu, Z. R., Li, Q. S., He, Y. C., & Chan, P. W. (2015). Gust factors for tropical cyclone, monsoon and thunderstorm winds. Journal of Wind Engineering and Industrial Aerodynamics, 142, 1-14.

Simiu, E. (2011). Design of buildings for wind: A guide for ASCE 7-10 standard users and designers of special structures. John Wiley & Sons.

Smith, J. M. (1991). Wind-wave generation on restricted fetches (No. CERC-MP-91-2). COASTAL ENGINEERING RESEARCH CENTER VICKSBURG MS.

Sobey, R. J., Harper, B. A., & Stark, K. P. (1977). Numerical simulation of tropical cyclone storm surge. James Cook University of North Queensland, Department of Civil & Systems Engineering. Also,

Sorensen, R. M. (2005). Basic coastal engineering (Vol. 10). Springer Science & Business Media.

Stansberg, C. T., Contento, G., Hong, S. W., Irani, M., Ishida, S., & Mercier, R. (2002). The specialist committee on waves final report and recommendations to the 23rd ITTC. In Proceedings of the 23rd ITTC (Vol. 2, pp. 505-551).

Sverdrup, H. U., & Munk, W. H. (1947). Wind, sea, and swell: theory of relations for forecasting. U.S. Navy Department, Hydrographic Office, Publication No. 601, 44 pp.

Tamura, Y., & Kareem, A. (Eds.). (2013). Advanced structural wind engineering. Springer Science & Business Media.

Thomas, T. J., & Dwarakish, G. S. (2015). Numerical wave modelling–A review. Aquatic Procedia, 4, 443-448.

Thompson, E. F., and Vincent, C. L. (1983). Prediction of wave height in shallow water. Proceedings of Coastal Structures 1983, American Society of Civil Engineers, 1000–1008.

Toba, Y. (1973). Local balance in the air-sea boundary processes. Journal of Oceanography, 29(5), 209-220.

Tolman, H. L. (2014), User manual and system documentation of WAVEWATCH III version 4.18. NOAA / NWS / NCEP / MMAB Technical Note 316, 194 pp.

Torsethaugen, K., & Haver, S. (2004). Simplified double peak spectral model for ocean waves. In The Fourteenth International Offshore and Polar Engineering Conference. International Society of Offshore and Polar Engineers.

Tucker, M. J. (1994). Nearshore waveheight during storms. Coastal Engineering, 24(1-2), 111-136.

U.S. Army Corps of Engineers (2015). Coastal Engineering Manual. Engineer Manual 1110-2-1100, Washington, D.C.: U.S. Army Corps of Engineers.

Valamanesh, V., Myers, A. T., Arwade, S. R., Hajjar, J. F., Hines, E., & Pang, W. (2016). Wind-wave prediction equations for probabilistic offshore hurricane hazard analysis. Natural Hazards, 83(1), 541-562.

Verkaik, J. W. (2000). Evaluation of two gustiness models for exposure correction calculations. Journal of Applied Meteorology, 39(9), 1613-1626.

Vickery, P. J., & Skerlj, P. F. (2005). Hurricane gust factors revisited. Journal of Structural Engineering, 131(5), 825-832.

Vincenty, T. (1975). Direct and inverse solutions of geodesics on the ellipsoid with application of nested equations. Survey review, 23(176), 88-93.

Wei, C. C. (2015). Forecasting surface wind speeds over offshore islands near Taiwan during tropical cyclones: Comparisons of data-driven algorithms and parametric wind representations. Journal of Geophysical Research: Atmospheres, 120(5), 1826-1847.

Wiberg, P. L., & Sherwood, C. R. (2008). Calculating wave-generated bottom orbital velocities from surface-wave parameters. Computers & Geosciences, 34(10), 1243-1262.

Wieringa, J. (1973). Gust factors over open water and built-up country. Boundary-Layer Meteorology, 3(4), 424-441.

Wilson, B. W. (1965). Numerical prediction of ocean waves in the North Atlantic for December, 1959. Ocean Dynamics, 18(3), 114-130.

World Meteorological Organization. Tropical Cyclone Programme, & Holland, G. J. (2015). Global guide to tropical cyclone forecasting. Secretariat of the World Meteorological Organization.

Wu, J. (1980). Wind-stress coefficients over sea surface near neutral conditions—A revisit. Journal of Physical Oceanography, 10(5), 727-740.

Yamartino, R. J. (1984). A comparison of several "single-pass" estimators of the standard deviation of wind direction. Journal of Climate and Applied Meteorology, 23(9), 1362-1366.

Young, I. R., & Verhagen, L. A. (1996a). The growth of fetch limited waves in water of finite depth. Part 1. Total energy and peak frequency. Coastal Engineering, 29(1-2), 47-78.

Young, I. R., & Verhagen, L. A. (1996b). The growth of fetch limited waves in water of finite depth. Part 2. Spectral evolution. Coastal Engineering, 29(1-2), 79-99.

Young, I. R., & Babanin, A. V. (2006). The form of the asymptotic depth-limited wind wave frequency spectrum. Journal of Geophysical Research: Oceans, 111(C6).

Young, I. R., & Vinoth, J. (2013). An "extended fetch" model for the spatial distribution of tropical cyclone wind–waves as observed by altimeter. Ocean Engineering, 70, 14-24.

Young, I. R. (2017). A Review of Parametric Descriptions of Tropical Cyclone Wind-Wave Generation. Atmosphere, 8(10), 194.

Zijlema, M., Stelling, G., & Smit, P. (2011). SWASH: An operational public domain code for simulating wave fields and rapidly varied flows in coastal waters. Coastal Engineering, 58(10), 992-1012.

Index

Made in United States
Orlando, FL
27 October 2025

71490218R00168